WORKBOOK OF TEST CASES FOR
VAPOR CLOUD SOURCE
DISPERSION MODELS

HARCROS
CHEMICAL GROUP

Presented to

with the compliments of

HARCROS CHEMICALS

WORKBOOK OF TEST CASES FOR
VAPOR CLOUD SOURCE DISPERSION MODELS

CENTER FOR CHEMICAL PROCESS SAFETY
of the
American Institute of Chemical Engineers
345 East 47th Street, New York, NY 10017

Library of Congress Cataloging-in-Publication Data

Workbook of test cases for vapor cloud dispersion models.
     p.   cm.
  Bibliography: p.
  Includes indexes.
  ISBN 0–8169–0455–3
  1. Atmospheric diffusion—Mathematical models. 2. Hazardous
substances—Environmental aspects—Mathematical models. 3. Vapors—
Mathematical models. I. American Institute of Chemical Engineers.
Center for Chemical Process Safety.
QC880.4.D44W67   1989
551.5′153—dc19   88–36745
                                                   CIP

**This book is available at a special discount when ordered in bulk quantities. For
information, contact the Center for Chemical Process Safety at the address given
above.**

Contents

WORKBOOK OF TEST CASES FOR
Vapor Cloud Source Dispersion Models

By

Steven R. Hanna
and
David Strimaitis

FOR

CENTER FOR CHEMICAL PROCESS SAFETY
of the
American Institute of Chemical Engineers

Acknowledgments

The AIChE wishes to thank the members of the Technical Steering Committee of the Center for Chemical Process Safety for their advice and support. Under the auspices of the Technical Steering Committee, the Vapor Cloud Committee of the CCPS provided guidance in this work.

The committee includes the following individuals: Thomas Carmody and Sandy Schreiber (AIChE staff members), Rudolph Diener, Chairman (Exxon Research and Engineering Co.), Gene Lee (Air Products and Chemicals), William Hague (Allied-Signal Corp.), Doug Blewitt (Amoco Corp.), Joe Tikvart (EPA), Jerry Schroy (Monsanto Corp.), Ronald Myers and Ronald Lantzy (Rohm & Haas), Gib Jersey (Mobil Research and Development), James Moser (retired from Shell Development Co.), B. Griffith Holmes (Westinghouse Electric Corp.), David Winegardner (Dow Chemical Co.), P. R. Jann (DuPont Co.), James Johnston (Merck & Co.), Arnold Marsden, Jr. (Shell Development Co.), and David McCready and Jerry R. Foster (Union Carbide Corp.).

The principal authors of the *Workbook of Test Cases for Vapor Cloud Source Dispersion Models* were Steven Hanna and David Strimaitis (Sigma Research Corp., Westford, Massachusetts).

Nomenclature

a	Cross-sectional area of tank (m^2)
A	Area of opening (m^2)
c_p	Specific heat at constant pressure (J kg^{-1} °K^{-1})
c_{pl}	Specific heat of liquid (J kg^{-1} °K^{-1})
c_v	Specific heat at constant volume (J kg^{-1} °K^{-1})
C	Concentration (kg m^{-3}, vppm)
C_0	Initial concentration (kg m^{-3}, vppm)
C_D	Discharge coefficient
C_{max}	Maximum ground level concentration (kg m^{-3}, vppm)
d	Effective pool diameter (m)
D	Diameter of release opening or pipe (m)
D_m	Molecular diffusivity of vapor in air (m^2 s^{-1})
f	Fraction flashed
F	Factor applied to two-phase releases
g	Acceleration due to gravity (9.8 m s^{-2})
h	Height of plume centerline (m)
H	Height of liquid above puncture (m)
k_g	Mass transfer coefficient (m s^{-1})
k_m	Kinematic viscosity of air (m^2 s^{-1})
L	Monin–Obukhov length (m)
L_e	Default pipe length of 0.1 m
L_p	Length of pipe from break to first valve (m)
mf	Mole fraction
M	Molecular weight of air (kg/kg-mol)
M_e	Effective molecular weight (kg/kg-mol)
M_g	Molecular weight of gas (kg/kg-mol)
M_0	Momentum flux (m^4 s^{-2})
N_{Sh}	Sherwood number
p	Pressure (N m^{-2})
p_a	Ambient pressure (N m^{-2})
p_s	Storage pressure (N m^{-2})
p_{vp}	Vapor pressure at storage pressure (N/m^2)
Ri_0	Initial Richardson number
Q	Continuous source strength (mass per unit time)
Q_e	Evaporative emission rate (mass per unit time)
Q_f	Mass of liquid that flashes (mass per unit time)
Q_g	Continuous source strength of gas (mass per unit time)
Q_l	Liquid mass emission rate (mass per unit time)
Q_t	Two-phase emission rate (mass per unit time)
r	Radius of cloud (m)

ix

r_0 Initial radius of cloud or plume (m)
R Gas constant (J kg^{-1} °K^{-1})
R^* Universal gas constant (8314.36 J kg-mol^{-1} °K^{-1})
t Time (s)
t_a Averaging time (s)
T Temperature (°K)
T_a Ambient temperature (°K)
T_b Normal boiling point of liquid (°K)
T_p Temperature of pool (°K)
T_s Storage temperature (°K)
u Wind speed (m s^{-1})
u_* Friction velocity (m s^{-1})
V Volume flow rate (m^3 s^{-1})
V_m Molar volume (m^3)
w_0 Initial plume vertical speed (m s^{-1})
W Plume width (m)
x Along-wind distance (m)
x_g Distance to plume touchdown (m)
x_0 Position of plume center in x direction (m)
x_{DO} DEGADIS and HMP solutions match at this distance (m)
x_{SO} SLAB and HMP solutions match at this distance (m)
x_{vy}, x_{vz} Virtual distances (m)
y Lateral distance from plume center; distance from source (m)
y_0 Lateral position of plume center (m)
z Vertical distance above ground (m)
z_0 Roughness length (m)
β Constant in DEGADIS σ_y formula
γ Gas specific heat ratio $= c_p/c_v$
δ Constant in DEGADIS σ_y formula
Δh Plume rise (m)
Λ Latent heat of vaporization (J kg^{-1})
ρ_a Ambient density (kg m^{-3})
ρ_g Gas density (kg m^{-3})
ρ_l Liquid density (kg m^{-3})
ρ_0 Cloud density (kg m^{-3})
σ_x Along-wind dispersion parameter (m)
σ_y Lateral dispersion parameter (m)
σ_z Vertical dispersion parameter (m)

Subscripts

a   Ambient conditions; averaging time
e   Effective value; refers to evaporative value
g   Refers to gas
l   Refers to liquid
0   Initial value
p   Pool

s	Storage value; saturation
t	Two phase
v	Virtual
x	Along-wind
y	Cross-wind
z	Vertical

Superscript

*	Refers to universal gas constant

1
Introduction and Description of Release Scenarios

There are many vapor cloud dispersion models available, with a great variation in capabilities and accuracies. The number of models changes almost daily as industries proceed with estimates of hazard zones from accidental releases of toxic or hazardous materials. In 1987 the Center for Chemical Process Safety published the *Guidelines for Use of Vapor Cloud Dispersion Models* (Hanna and Drivas, 1987), which reviewed the physical principles in source emission and transport and dispersion models, and summarized the capabilities of 32 different models. However, there were no examples of procedures to follow in the case of a specific chemical release. The purpose of the present workbook is to step through the calculations one might typically make for five specific release scenarios, using publicly available formulas and computer models. It will be seen that it is generally necessary to patch two or more models together to better simulate these realistic scenarios. The use of any particular modeling method in this workbook does not constitute endorsement by the CCPS or the authors.

The CCPS Vapor Cloud Dispersion committee pooled their experience to generate the list of five release scenarios pictured in Figure 1-1 and briefly described in Table 1-1. A wide range of release types is represented, including an elevated continuous release of dense normal butane gas (Scenario 1), a release of liquid ammonia from a pressurized tank (Scenario 2), a low-level jet release of carbon monoxide gas (Scenario 3), a release of liquid chlorine from a pressurized tank (Scenario 4), and a release of liquid acetone into a diked area (Scenario 5). In the ammonia and chlorine scenarios, some of the liquid flashes into a gas, some mass spills on the ground, and some may be entrained in the jet as an aerosol. In the acetone scenario, the liquid all spills into the diked area and evaporates. In all cases the variation of maximum ground level concentration of the gas as a function of downwind distance is desired. Surface roughness lengths (z_0) of 0.03 and 0.3 m are to be used in models that can accept this information. These correspond to a mowed grass surface and a suburban or industrial surface, respectively (note that local building effects are not considered). Calculations are to be made for light wind stable conditions (class F, with wind speed 2 m/s) and

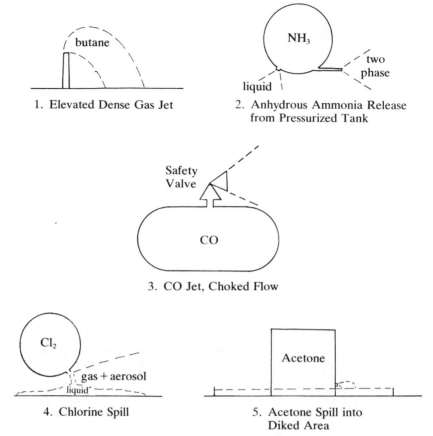

Figure 1-1. *Diagrams of five release scenarios.*

neutral conditions (class D, with wind speeds 2 and 5 m/s). This range of input conditions is used to illustrate how sensitive the models are to typical variations in wind speed, stability, and roughness length. The effects of variations from 0 to 50% in entrained aerosol fractions are investigated in Scenario 4 (chlorine). The state of the art does not allow prediction of exact aerosol fraction.

In Scenarios 1 through 4, methods of handling the transition from initial jet model to dense gas slumping model are discussed. The models themselves provide little guidance on this problem, and it is seen that there are often several ways to approach it.

The first four scenarios were selected so that the excess density in the puff or plume would be significant, requiring the use of a dense gas model. In any other practical problem, the source Richardson number, Ri_0, should be calculated to determine whether it is necessary to use a dense gas model. If not, then a simple Gaussian model can be applied. Hanna and Drivas (1987) show that there are a number of slightly different ways to define Ri_0, but a generally acceptable definition for continuous plumes is $Ri_0 = gw_0[(\rho_p - \rho_a)/\rho_a]D/u_*^2u$. In this expression, w_0 and D are the initial plume speed and diameter, ρ_p and ρ_a are initial plume and ambient density, u is wind speed, and u_* is ambient friction

TABLE 1-1
Overview of RELEASE SCENARIOS

Scenario description	Modeling concern	Procedure/model selection	Release conditions[a]
1. Elevated release of normal butane nC_4 via 0.75-m (i.d.) vent stack at 5-m elevation	Plume touchdown distance and concentration at touchdown. Maximum downwind centerline concentration for dilution to LFL and 25% LFL (instantaneous concentration basis; taken to be 10 s)	Hoot/Meroney/Peterka model for plume touchdown. SLAB, DEGADIS, and Gaussian models for downwind transport	nC_4 at 15 kg/s and 265°K. Ambient at 30°C, 50% RH. Release duration = 30 min
2. Ground level release of liquid anhydrous ammonia from ambient temperature storage (40 mT) by catastrophic failure in 1.5-in. connecting orifice and also 1.5-in. line—~150 in. from vessel	Maximum downwind centerline concentration vs distance, 15- and 30-min averaging times	Bernoulli equation for orifice discharge. Fauske equation for critical flow from line discharge. DEGADIS and Gaussian models for downwind transport	Ammonia stored at 8.9 bar and 25°C. Ambient at 25°C, 80% RH. Release duration = 30 min. Assume complete vapor/aerosol formation, i.e., no liquid pool
3. Low-level release of high-pressure CO from relief valve (10.2 cm diameter initial jet)	Maximum downwind centerline concentration vs distance for 1- and 15-min averaging times	Critical flow equation for source term. Briggs and Ooms models for jet entrainment. Gaussian model for downwind transport	RV discharge horizontally at 21°C. Burst pressure of 2230 psig. Release duration = 5 min
4. Ground level release of liquid Cl_2 from 0.3-in. i.d. valve failure on a 1-ton cylinder. Consider variable aerosol fractions (0 and 50%). Include secondary pool source term	Maximum downwind centerline concentration vs distance for 1- and 15-min averaging times	Bernoulli equation for release rate. SPILLS model for evaporation. SLAB, DEGADIS, and CAMEO models for downwind transport	Liquid Cl_2 saturated at 30°C. Spill occurs on wet soil undiked surface. Ambient at 30°C, 80% RH. Release duration—until cylinder is empty
5. Ground level release of acetone from 2-in. nozzle connection from 25 ft diameter × 25 ft H storage within a concrete diked area of height 3 ft and diameter 72 ft	Maximum downwind centerline concentration vs distance for 1- and 30-min averaging times	SPILLS model for evaporation rate. SPILLS and DEGADIS models for downwind transport	Acetone stored at 30°C and ambient pressure. Ambient at 30°C, 50% RH. Release duration = 30 min

[a]For all scenarios, calculations are made for a surface roughness of 0.03 and 0.30 m, for stability class F (with wind speed 2 m/s), and for stability class D (with wind speeds 2 and 5 m/s).

3

velocity (roughly equal to $0.065u$). If $Ri_0 \geqslant 10$ then one of the dense gas models should be used. Otherwise the Gaussian model or one of the standard EPA models for buoyant or neutrally buoyant plumes can be used.

These scenarios were also selected so that the source emission rate would be relatively constant over a period of at least several minutes, thus justifying the use of a steady-state or continuous plume model. If the emission rate were highly time dependent or nearly instantaneous, it would be necessary to use a transient model or a puff model. Such models exist for dense gas releases, but are generally less well tested.

A discussion of the equations and models to be used is given in Chapter 2 and the five scenarios are worked through in Chapter 3. Listings of typical computer input and output are given in the Appendix.

2

Overview of Equations and Models

A vapor cloud dispersion model is a series of procedures for calculating concentrations of a gas downwind of the release point. In some cases procedures for estimating source emissions are also included. For convenience, most models are organized into a computer code. Anyone who studies one of these codes discovers that many of the statements are used for organizing input and output data and doing general bookkeeping within the code. The set of equations that is used is usually relatively small. Consequently in most cases it is possible to do a "hand-check" on various sections of the computer model by carrying out hand calculations of the individual equations. Both hand calculations and computer model runs are carried out in this workbook. The equations and the computer models that are used are briefly described below, beginning with source emissions models, and moving to transport and dispersion models. Formulas for calculating the transition from one model to another are given.

2.1 Source Emissions Calculations

Most hazardous gas releases to the atmosphere are the result of pipeline or tank ruptures, resulting in gas or liquid jets and/or liquid spills onto the ground, with subsequent evaporation. In many cases it is possible to calculate source emissions using simple analytical formulas, as summarized in the following sections.

2.1.1 *Liquid or Two-Phase Jet from Pressurized Tank*

Some potential scenarios for jet releases from pressurized tanks are illustrated in Figure 2-1 for release types A through E discussed below. If the hole is right at the tank and is below the liquid level in the tank, the mass flux through the opening should be calculated as all liquid flow, with any flashing occurring in the discharge jet region. If the pipe breaks at a distance of about 0.1 m or more from

5

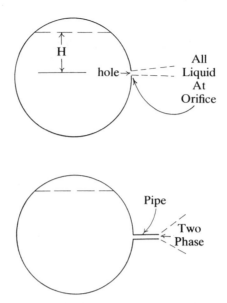

Figure 2-1. *Source scenarios considered under Section 2.1.1 on liquid or two-phase jets from pressurized tanks. The upper part of the figure refers to release types A and B and the lower part refers to release types C, D, and E, as described in this section.*

the tank, then the mass flux should be calculated assuming two-phase flow in the pipe (Fauske and Epstein, 1987).

RELEASE TYPE A. "CLASSICAL" APPROACH
The Bernoulli equation is recommended for calculating the mass flux of liquid from a hole in a pressurized tank. Neglecting frictional losses and assuming that any flashing that might occur on depressurization occurs downstream of the opening (i.e., all liquid flow through the hole), then the liquid flow can be calculated per Perry and Green (1984):

$$Q_1 = C_D A \rho_1 (2\Delta p / \rho_1 + 2gH)^{1/2} \qquad (2\text{-}1)$$

where Q_1 is mass emission rate (kg/s)
　　　A is area of pipe or tank opening (m^2)
　　　ρ_1 is density of liquid in tank (kg/m^3)
　　　Δp is the difference between the pressure of the liquid in the tank and the ambient pressure (N/m^2)
　　　g is the acceleration of gravity (9.8 m/s^2)
　　　H is the depth of liquid in the tank above the hole (m)
　　　C_D is the discharge coefficient, assumed to equal 0.6.

For the case in which the pressure in the tank remains equal to the ambient pressure, $\Delta p = 0$, the mass release rate is simply proportional to the square root of the height of liquid in the tank above the hole. Assuming that $\Delta p = 0$ and mass is conserved, we derive the relation

$$Q_l(t) = [0.6 A \rho_l (2 g H_0)^{1/2} - t(0.6 A)^2 \rho_l g / a] \tag{2-2}$$

where H_0 is the initial height of liquid in the tank at time $t = 0$, and a is the cross-sectional area of the tank (in m^2). It is assumed that the cross-sectional area of the tank is invariant with height (e.g., an upright cylinder). Equation (2-2) implies that the tank will be empty at a time $(1.67 a / A) (2 H_0 / g)^{1/2}$.

RELEASE TYPE B. FAUSKE–EPSTEIN EQUATION FOR SUBCOOLED LIQUIDS
Fauske and Epstein (1987) found that if the stored liquid is externally pressurized (e.g., via use of a nitrogen pad), the vapor pressure is greater than one atmosphere, and the storage temperature is below the saturation temperature associated with storage pressure, then Eq. (2-1) must be modified. The vapor pressure at storage temperature should be used in place of the ambient pressure:

$$Q_l = C_D A \rho_l \{[2(p_s - p_{vp}) / \rho_l] + 2 g H\}^{1/2} \tag{2-3}$$

where p_s is storage pressure (N/m^2)
$\quad p_{vp}$ is vapor pressure at storage temperature (N/m^2).

The release rate calculated by this equation can be substantially less than that calculated by Eq. (2-1).

RELEASE TYPE C. FAUSKE–EPSTEIN EQUATION FOR SATURATED LIQUIDS
If the liquid is at saturation (i.e., $p_s = p_{vp}$) and equilibrium two-phase choked flow is established (i.e., pipe length > 0.1 m), then Fauske and Epstein (1987) recommend the following equation:

$$Q_t = [A \Lambda / (\rho_g^{-1} - \rho_l^{-1})](T_s c_{pl})^{-1/2} \tag{2-4}$$

where Q_t is total two-phase emission mass rate (kg/s)
$\quad \Lambda$ is latent heat of vaporization (J/kg)
$\quad \rho_g$ is gas density at storage pressure (kg/m^3)
$\quad T_s$ is the storage temperature (°K)
$\quad c_{pl}$ is specific heat of liquid (J/kg/°K).

This equation applies only if the following condition is met:

$$x < p_s (\rho_g^{-1} - \rho_l^{-1})(T_s c_{pl}) / \Lambda^2 \tag{2-5}$$

where x is the weight fraction of vapor after depressurizing to atmospheric pressure. If this condition is not met, then a more complicated numerical model is necessary to calculate the emission rate (Fauske and Epstein, 1987).

RELEASE TYPE D. FRICTIONAL LOSSES FOR SATURATED LIQUIDS
For long pipe lengths the mass emission rate, Q_t, in Eq. (2-4) should be multiplied by a factor, F, that accounts for frictional losses. Suggested values for F are given in Table 2-1, where L_p and D are pipe length and diameter, respectively (Fauske and Epstein, 1987). These estimates of F are based on

TABLE 2-1
Variation of Factor F with Ratio L_p/D^a

L_p/D	F
0	1
50	0.85
100	0.75
200	0.65
400	0.55

^{a}From Fauske and Epstein (1987).

limited experiments and theoretical calculations and may not be universally valid for all conditions. The reader is referred to Fauske and Epstein (1987) for further discussions of the limitations.

RELEASE TYPE E. NONEQUILIBRIUM TWO-PHASE FLOW
According to Fauske and Epstein (1987), for pipe lengths, L_p, less than 0.1 m, equilibrium two-phase choked flow will not be established, and the mass emission rate, Q_t, in Eq. (2-4) should be multiplied by the factor $N^{-1/2}$, where N is given by

$$N = \Lambda^2/[2(\Delta p)\rho_l C_D^2(\rho_g^{-1} - \rho_l^{-1})^2 T_s c_{pl}] + L_p/L_e \qquad (2\text{-}6)$$

where L_p is pipe length to opening (m) and $L_e = 0.1$ m. In the limit, as pipe length L_p approaches 0.0, the solution approaches the Bernoulli equation (2-1).

FRACTION FLASHED
For superheated liquids (i.e., stored at temperature above the normal boiling point), a fraction, f, of the liquid emission is "flashed" to vapor as the pressure is reduced to ambient. This fraction is approximated from the thermodynamic relationship

$$f = c_{pl}\Delta T/\Lambda \qquad \Delta T = (T_s - T_b)\,(°K) \qquad (2\text{-}7)$$

where T_s is storage temperature and T_b is normal boiling point temperature. The flashing is accompanied by a rapid expansion of the jet as the density of the chemical is reduced by about three orders of magnitude. Two-phase jets are observed to cover angles up to about 100° near the nozzle or other opening. This rapid expansion, along with the turbulence generated by a high momentum jet, can cause the remaining liquid to be broken up into an aerosol (small drops) that is carried downwind. In many cases, estimating the fraction of liquid entrained as an aerosol is highly uncertain.

2.1.2 Gas Jet from Pressurized Tank

If the material is stored as a gas and the pressure, p_s, in a tank is greater than about two times the pressure, p_a, of the ambient air, then critical flow will exist.

The following criterion is well-known (Perry and Green, 1984):

$$\text{Critical flow if } p_s/p_a \geq [(\gamma + 1)/2]^{\gamma/(\gamma-1)} \tag{2-8}$$

where $\gamma = c_p/c_v$.

If criterion (2-8) is satisfied, the emission rate is calculated with the formula

$$Q = C_D\{(\gamma M/R^* T_s)[2/(\gamma + 1)]^{(\gamma+1)/(\gamma-1)}\}^{0.5} \tag{2-9}$$

where M is the molecular weight, A is the area of the orifice, and T_s is the tank temperature. If the criterion is not satisfied, then the flow becomes subcritical and the following formula should be used:

$$Q = C_D A\{2\rho_g p_s[\gamma/(\gamma - 1)][(p_a/p_s)^{2/\gamma} - (p_a/p_s)^{(\gamma+1)/\gamma}]\}^{1/2} \tag{2-10}$$

where ρ_g and p_s are the gas density and pressure in the tank. The discharge coefficient, C_D, is usually assumed to be 0.8 for choked flow (Perry and Green, 1984).

2.1.3 *Slowly Evaporating Pool*

In many cases the hazardous material does not evaporate before it hits the ground surface, and it is necessary to model the evaporation from a surface pool. The SPILLS model (Fleischer, 1980) is often applied to estimate the mass emission rate from pools of evaporating single-component liquids. If the rate of evaporation is light to moderate (i.e., the pool temperature is within a few degrees of ambient and the liquid does not boil), and the liquid is well mixed, then an empirical formula for slowly evaporating pools can be applied.

$$Q_e = k_g A p_{vp} M/R^* T_p \tag{2-11}$$

where Q_e is the evaporative emission rate (kg/s)

    A is the pool area (m²)

    p_{vp} is the vapor pressure (N/m²)

    M is the molecular weight (kg/kg-mol)

    R^* is the gas constant (J/mol/°K)

    T_p is the pool temperature (°K)

The parameter k_g is the mass transfer coefficient (m/s), given by the formula

$$k_g = D_m N_{Sh}/d \tag{2-12}$$

where D_m is the molecular diffusivity of the vapor in air (m²/s)

    d is the effective pool diameter (m)

    N_{Sh} is the Sherwood number, given by the correlation

$$N_{Sh} = 0.037(k_m/D_m)^{1/3}[(ud/k_m)^{0.8} - 15200] \tag{2-13}$$

where k_m is the kinematic viscosity of the air (m^2/s)

    u is the wind speed at 10 m over the pool (m/s).

It is assumed that the air flow over the pool is turbulent. The reader is cautioned that the values of D_m are not easily found for more than a few basic chemicals. They are "hard-wired" in the SPILLS model for some chemicals, but for other chemicals the user must develop these numbers based on principles of physical chemistry (Reid *et al.*, 1977; Lyman *et al.*, 1982).

In the case of boiling pools, the equations become more complicated and the reader is referred to Fleischer (1980) for a complete discussion. For example, if the boiling is sufficiently rapid, the mass emission rate is limited by the rate of heat transfer from the ground to the pool. If the ground freezes, the heat transfer can be significantly reduced, and the evaporation from the pool is also greatly reduced.

2.2 Momentum Jet and HMP Models

In Section 2.1, formulas were provided for calculating the emission rate of gas jets from pipes or tanks. Given the momentum flux of gases released from a source into a crosswind of speed, u, the distance traveled by the momentum jet before it is turned by the airflow can be calculated using a formula suggested by Briggs (1975):

$$\Delta h = 4.8 M_0^{1/2}/u \qquad\qquad (2\text{-}14)$$

where u is the wind speed (m/s), which is assumed to be blowing perpendicular to the direction of the jet axis, and M_0 is the initial momentum flux (m^4/s^2) divided by ρ. For a jet with uniform initial velocity w_0 from a pipe with radius R_0, $M_0 = w_0^2 R_0^2$. The subsequent growth of the plume radius can be estimated from the formula

$$R = R_0 + (0.4 + 1.2 u/w_0)z \qquad\qquad (2\text{-}15)$$

where z is crosswind distance from the source, which can be up, down, or sideways, as long as the jet is perpendicular to the windflow and there are no obstructions to the jet.

When the momentum jet is heavier than air, the above equations are not appropriate and an alternate technique is required. One method is based on a series of laboratory experiments on elevated dense gas jets that was performed by Hoot, Meroney, and Peterka (HMP, 1973), who derived a set of empirical formulas for the maximum plume rise of vertically oriented dense jets, the downwind distance at which the dense plume strikes the ground, and the maximum ground level concentration at that point and at points further downwind. The Ooms *et al.* (1974) model is also appropriate for this scenario and agrees quite closely with the HMP model (see Section 2.5.1). These formulas have been derived from experiments where the initial Richardson number, Ri_0, exceeded about 30. The model equations are given by Hanna and Drivas (1987) on pp. 53–58, and are repeated below (the reader will note some changes, since corrections were made to two errors in Hanna and Drivas).

The following equations are used to calculate the maximum rise above the stack, Δh, the downwind distance to plume touchdown, x_g, and the concentration ratio C/C_0 at plume touchdown:

$$\Delta h/2R_0 = 1.32(w_0/u)^{1/3}(\rho_0/\rho_a)^{1/3}\{w_0^2\rho_0/[2R_0g(\rho_0-\rho_a)]\}^{1/3} \tag{2-16}$$

$$x_g/2R_0 = \{w_0 u\, \rho_0/[2R_0g(\rho_0-\rho_a)]\} +$$
$$0.56\{(\Delta h/2R_0)^3[(2+h_s/\Delta h)^3-1]\}^{1/2}\{u^3\rho_a[2R_0gw_0(\rho_0-\rho_a)]\}^{1/2} \tag{2-17}$$

$$C/C_0 = 2.43(w_0/u)[(h_s+2\Delta h)/2R_0]^{-1.95} \quad \text{(at the point } x_g) \tag{2-18}$$

where w_0, ρ_0, R_0, and C_0 are the initial plume speed, density, radius, and mass concentration. The variables u and ρ_a are the ambient wind speed and density, respectively.

2.3 Gaussian Formula for Plume Dispersion

Once the density of the gas in the plume is close to that of the ambient air, it is dispersed by ambient turbulence. The Gaussian formula is often the basis for calculating dispersion in these conditions, and is the foundation for most models recommended by the EPA for application to neutrally or positively buoyant releases of gases.

$$C = [Q/(2\pi\sigma_y\sigma_z u)]\exp[-(y-y_0)^2/2\sigma_y^2]\{\exp[-(h-z)^2/2\sigma_z^2]+$$
$$\exp[-(-h-z)^2/2\sigma_z^2]\} \tag{2-19}$$

where C is the concentration (in kg/m³)
  Q is the continuous source emission rate (in kg/s)
  σ_y is the lateral dispersion coefficient, or standard deviation of the lateral concentration distribution (m)
  σ_z is the vertical dispersion coefficient, or standard deviation of the vertical concentration distribution (m)
  u is the wind speed at plume height (m/s)
  h is the elevation of the plume centerline above the ground (m)
  y is the lateral crosswind position of interest (m)
  y_0 is the lateral crosswind position of plume centerline (m)
  z is the height above ground of the position of interest (m).

It is assumed that reflection of the plume at the ground occurs, which is accounted for by the last exponential term involving $(-h-z)$.

Equation (2-19) is valid for a continuous source. If there is an instantaneous source of strength Q_i (kg), then dispersion, σ_x, in the along-wind direction (x) must also be accounted for, and the equation becomes

$$C = [Q_i/(2^{3/2}\pi^{3/2}\sigma_x\sigma_y\sigma_z)]\exp[-(x-x_0)^2/2\sigma_x^2]$$
$$\times \exp[-(y-y_0)^2/2\sigma_y^2]\{\exp[-(h-z)^2/2\sigma_z^2]+\exp[-(-h-z)^2/2\sigma_z^2]\} \qquad (2\text{-}20)$$

This equation gives the instantaneous concentration for a puff centered at (x_0, y_0, z). Distance can be converted to time via the relation $x_0 = ut_0$, where t_0 is time of travel. If an average over a time duration from t_1 to t_2 is required, Eq. (2-20) must be integrated over that range. Most models assume $\sigma_x = \sigma_y$, although some models include the effects of wind shear (i.e., wind increase with height) in the σ_x calculation, which may cause additional dilution of the cloud (Hanna and Drivas, 1987). The dispersion coefficients, σ_x, σ_y, and σ_z are usually given as empirical functions of downwind distance, x.

Briggs (1973) suggests formulas for σ_y and σ_z for continuous plumes for averaging times of about 10 min as a function of downwind distance, x, in rural and urban conditions, as listed in Table 2-2. All of these formulas can be written in the form $\sigma = ax(1 + bx)^p$, where a, b, and p are parameters that are functions of stability class.

TABLE 2-2
Briggs's (1973) Formulas for Dispersion Coefficients for Continuous, Neutrally Buoyant Plumes

Stability class	$\sigma_y(m)$	$\sigma_z(m)$
Rural		
A	$0.22x(1+0.0001x)^{-1/2}$	$0.20x$
B	$0.16x(1+0.0001x)^{-1/2}$	$0.12x$
C	$0.11x(1+0.0001x)^{-1/2}$	$0.08x(1+0.0002x)^{-1/2}$
D	$0.08x(1+0.0001x)^{-1/2}$	$0.06x(1+0.0015x)^{-1/2}$
E	$0.06x(1+0.0001x)^{-1/2}$	$0.03x(1+0.0003x)^{-1}$
F	$0.04x(1+0.0001x)^{-1/2}$	$0.016x(1+0.0003x)^{-1}$
Urban		
A–B	$0.32x(1+0.0004x)^{-1/2}$	$0.24x(1+0.001x)^{1/2}$
C	$0.22x(1+0.0004x)^{-1/2}$	$0.20x$
D	$0.16x(1+0.0004x)^{-1/2}$	$0.14x(1+0.0003x)^{-1/2}$
E–F	$0.11x(1+0.0004x)^{-1/2}$	$0.08x(1+0.0015x)^{-1/2}$

Stability class A refers to unstable light-wind sunny conditions, D to neutral conditions (windy or cloudy), and F to stable light-wind clear nighttime conditions. These formulas are valid for an averaging time of about 10 min. If the actual averaging time, t_a, is much different from 10 min, the following empirical correction to σ_y can be used (Hanna *et al.*, 1982):

$$\sigma_y(t_a)/\sigma_y(10\text{ min}) = [t_a/(10\text{ min})]^{0.2} \qquad (2\text{-}21)$$

The power 0.2 in Eq. (2-21) represents an average over several field data sets, and actually is a function of stability class, downwind distance, and other variables.

Equation (2-21) should not be used if the calculated $\sigma_y(t_a)$ drops below the known σ_y value for instantaneous puffs or plumes. In practice, σ_y should be the *maximum* of that given by Eq. (2-21) and the instantaneous value given by the following equations from Slade (p. 175 of *Meteorological and Atomic Energy*,

1968), where the leading constants for stability classes A, C, and E have been interpolated or extrapolated from the given constants for stability classes B, D, and F.

Stability class	A	B	C	D	E	F
σ_{y1}	$0.18x^{0.92}$	$0.14x^{0.92}$	$0.01x^{0.92}$	$0.06x^{0.92}$	$0.04x^{0.92}$	$0.02x^{0.92}$

All of the σ_y and σ_z formulas are based on the results of full-scale field experiments. The rural field experiments were conducted over short grassy surfaces with roughness lengths in the range from 0.01 to 0.1 m, and the urban experiments were conducted over typical mixed surfaces with roughness lengths on the order of 1 m. None of these experiments used to derive the formulas in Table 2-2 or the relation in Eq. (2-21) involved the release of tracer gases in sufficient quantity that dense gas effects were important.

Most of the dense gas computer codes contain equations equivalent to the rural σ_y and σ_z formulas given in Table 2-2, but none of these codes has the capability to simulate the urban formulas. These rural formulas are used as far-downwind asymptotes in the dense gas models. Furthermore, we emphasize that the models all use the continuous plume σ_y and σ_z formulas to also simulate instantaneous puffs.

2.4 Transition from One Model Type to Another

It is sometimes necessary to use one model at small downwind distances and another model at large distances. At the transition point, mass fluxes should be conserved, maximum concentrations should be continuous, and the plume dimensions should be conserved, if possible.

2.4.1 *Jet to Gaussian*

The Gaussian equation should be used only at downwind distances where the initial jet effects have become unimportant and the plume density has dropped to nearly ambient. Use of the Gaussian equation at distances less than 100 m can create additional uncertainty. The plumes and puffs resulting from releases of hazardous chemicals in the five scenarios described in Chapter 3 will initially be relatively dense, and many models use a relative density criterion for triggering the transition to the Gaussian formula. For example, if ρ_g and ρ_a are the gas density and ambient density, then dense gas effects cease to be important when the relative differences $(\rho_g - \rho_a)/\rho_a$ drop below a value in the range from 0.001 to 0.02. Different models use different relative density criteria, with DEGADIS at the high end of the range. The transition is usually accomplished by matching the Gaussian σ_y and σ_z with the distribution parameters (e.g., cloud height and width) resulting from the application of some dense gas model. Dense gas models use a variety of distribution assumptions, including the following three:

- Round plume or puff with diameter $2R$ and constant concentration, C.
- Box or slab with height h, width W, and constant concentration, C.
- Gaussian or other simple shape.

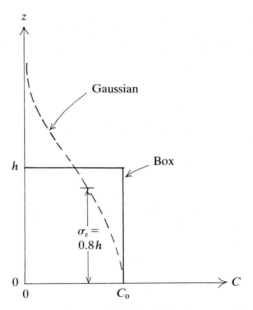

Figure 2-2. *Matching Gaussian with box or slab distributions at transition point. Both maximum concentration (C_0) and mass fluxes ($u \int_0^\infty C(z)\,dz = uC_0\sigma_z\sqrt{\pi/2}$ for Gaussian and $= uC_0h$ for box) are conserved, giving $\sigma_z = \sqrt{2/\pi}h = 0.80h$ for this example.*

For example, consider the box or slab distribution. To conserve mass and match the peak concentration for plumes or clouds resting on the ground, it is necessary that the product $\sigma_y\sigma_z$ equals hW/π, which can be accomplished if $\sigma_y = \sigma_z = (hW/\pi)^{1/2}$ (see Figure 2-2). But the actual downwind distance probably does not equal the virtual distance, x_v, that would give these values of σ_y and σ_z in Briggs's (1973) formulas listed in Table 2-2. In this case a virtual distance, x_{vy} or x_{vz}, is calculated, using the σ_y and σ_z formulas in Table 2-2:

For $\sigma_y = ax(1 + bx)^{-1/2}$ (All Rural and Urban Classes)

$$x_{vy} = (b\sigma_y^2 + \sqrt{b^2\sigma_y^4 + 4a^2\sigma_y^2})/2a^2 \tag{2-22}$$

For $\sigma_z = ax(1 + bx)^{-1/2}$ (Rural Class C and D and Urban Class D–F)

$$x_{vz} = (b\sigma_z^2 + \sqrt{b^2\sigma_z^4 + 4a^2\sigma_z^2})/2a^2$$

For $\sigma_z = ax$ (Rural Class A and B and Urban Class C)

$$x_{vz} = \sigma_z/a \tag{2-24}$$

For $\sigma_z = ax(1 + bx)^{-1}$ (Rural Class E and F)

$$x_{vz} = \sigma_z/(a - b\sigma_z) \tag{2-25}$$

For urban classes A and B, the cubic equation is rather complicated, and the approximation $x_{vz} = \sigma_z/0.24$ can be used. In the σ_y formulas the coefficient a should be multiplied by the averaging time factor given in Eq. (2-21) before calculating the virtual distances. Note that x_{vy} does not necessarily equal x_{vz}.

Also, Eq. (2-25) has no solution if σ_z is so large that $\sigma_z \geq a/b$ (100 m for rural class E and 53 m for rural class F). In any case, the new σ_y and σ_z values to be used in the Gaussian equation at a distance x_z from the transition point are those appropriate for distances x_{vy} and x_{vz}, respectively.

2.4.2 *Jet to SLAB or DEGADIS*

The transition of a dense-gas plume from an initial jet model to a dense-gas slumping model in either SLAB or DEGADIS relies on the notion that once the dense-gas plume reaches the surface, its evolution can be approximated by that of a plume from a ground level release with the same mass flux rate. Both SLAB and DEGADIS assume that the emissions are from a ground level area source. It is required that mass be conserved. Conservation of other fluxes such as enthalpy or momentum is not considered. To satisfy the condition that the peak concentration in the plume at touchdown match the concentration in the SLAB or DEGADIS model, the dense-gas source must be diluted by an arbitrary factor (i.e., the mole fraction of the dense gas released set less than 1), or the concentration must be matched at some distance downwind of the area source (requiring calculation of a virtual source distance).

One way to match concentrations as well as conserve mass in the SLAB model is to identify the distance at which the concentration at the centerline of the plume in SLAB matches the peak concentration in the jet at touchdown. This distance is then identified as the point at which the transition takes place. The area of the source for SLAB can be chosen to match the area of the jet at touchdown. Another way to match concentrations (not discussed further here) is to approximate the initial cloud in SLAB as a dilute mixture of dense gas and air. This procedure is not completely described in the SLAB model reports (e.g., Ermak and Chan, 1985) and requires some practice to refine the procedure.

DEGADIS will accept a mixture of a dense gas and air, so that both mass flow rate and concentration can be matched at the DEGADIS source. However, the model frequently produces additional dilution of the mixture at the source, depending on what area is specified for the source. The initial dilution can be removed by a sufficiently large choice of area. Or it can sometimes be countered by reducing the amount of air in the initial mixture released at the source. In either case, some iterations are required to get the initial concentrations from DEGADIS to agree (approximately) with the peak concentration in the jet at touchdown. Note that DEGADIS can also be run in the same way that SLAB is run, that is, by calculating the virtual distance at which the SLAB concentration matches the peak concentration in the jet at touchdown.

There are currently three versions of DEGADIS available. DEGADIS 1.4 and 1.5 do not treat jets and thus must employ the transition assumptions listed here. DEGADIS 2.0 does treat a jet and automatically calculates the transition within the model. The latter version was not available until this workbook was nearly completed.

When DEGADIS 1.4 or 1.5 is used, it must be run in isothermal mode so that only mixing of the diluted gas with air need be considered. The mixture is specified in the following way. Let *mf* be the mole fraction of the dense gas in the mixture at the transition point where DEGADIS is initialized. This gas has a molecular weight M_g, while that of air is M_a. The molecular weight of the initial mixture is given by

$$M = mf \times M_g + (1 - mf)M_a \qquad (2\text{-}26)$$

If V_m is the molar volume at the ambient temperature, then the initial density of the mixture is

$$\rho = M/V_m \qquad (2\text{-}27)$$

The emission rate (mass/unit time) must also be adjusted because the mixture contains $(1 - mf)M_a$ mass units of air for every $mf \times M_g$ mass units of dense gas. The mixture emission rate is

$$Q = Q_g M/(mf \times M_g) \qquad (2\text{-}28)$$

where Q_g is the emission rate of the dense gas in the original jet.

In this workbook, only the steady-state versions of the models are run. If the travel time exceeds the release time, some overestimation of concentrations may result. Users generally report that it is often difficult to obtain reasonable solutions with the transient versions of the models.

2.5 Computer Codes Used in Scenarios

A variety of computer codes were applied by the five scenarios to estimate source emissions and subsequent transport and dispersion. These models are briefly discussed here, and the reader is referred to the companion document by Hanna and Drivas (1987) or to the relevant users' guides for more detailed discussions. We caution the reader that many of these codes are constantly being modified. The versions that we use are identified, but the version that the reader obtains may be different.

2.5.1 *Ooms Model*

The Ooms model is a theoretical model intended for application to the same elevated dense-gas jet scenario as the HMP model (discussed under Section 2.2). An example of the form of the equations in this model (Ooms *et al.*, 1974) is given by Hanna and Drivas (1987) on p. 58. The Ooms model has recently been coded by Havens (1987) and is included either as part of the DEGADIS 2.0 model (runs on a VAX computer) or as an independent model (runs on a VAX computer or IBM PC).

2.5.2 *SPILLS Model*

The Shell SPILLS model (Fleischer, 1980) has been in use for many years and has been incorporated into several other models. It contains empirical formulas for calculating the evaporative emissions from liquid spills. Downwind transport and dispersion are estimated under the assumption that the cloud is neutrally buoyant. It is described by Hanna and Drivas (1987) on pp. 30, 38, 39, 114, 140, and 142, and is available from the National Technical Information Service.

2.5.3 *DEGADIS Model*

The development of the DEGADIS model was sponsored by the U.S. Coast Guard to calculate dense-gas slumping, transport, and dispersion (Havens and Spicer, 1985). It assumes that an area source emission rate is given. It is best applied to the near-field gravity slumping of a dense cloud. Hanna and Drivas (1987) discuss the DEGADIS model on pp. 36, 39, 104–107, and 141. We note that many revisions are currently under way. The version that we used is an IBM PC version of DEGADIS 1.4 that was converted by D. Blewitt of Amoco. Mr. Blewitt was able to match his predictions with the DEGADIS VAX model predictions for the test cases in the user's guide, plus several runs from the Burro and Goldfish field experiments. We corrected for averaging time using power law formulas contained in DEGADIS 1.5 [see Eq. (2-21)].

2.5.4 *SLAB Model*

The SLAB model (Ermak and Chan, 1985) is similar in range of application to the DEGADIS model. Both assume area source emissions and account for dense-gas slumping. Hanna and Drivas (1987), pp. 123 and 142, cover the SLAB model. The SLAB model is also being revised, and the version we used is dated December 1987.

2.5.5 *CAMEO/ALOHA Model*

The U.S. Department of Commerce has recently been distributing the CAMEO model, which is intended for use by local agencies in assessing the potential environmental impact of releases of hazardous chemicals. The model includes a large chemical data base, but its atmospheric dispersion algorithm (ALOHA) is very simple at the moment. ALOHA currently includes only the steady-state or instantaneous Gaussian model for nondense gasses, although there are plans to add more sophisticated algorithms by 1989. This model requires an Apple MacIntosh computer. Version 3.3 of CAMEO/ALOHA was used in Scenario 4.

3
Application of Models to Five Scenarios

In this section, the models are applied to the five release scenarios discussed in Chapter 1. All models used in these applications are publicly available and have been reviewed in Chapter 2. Readers are cautioned that some chemical engineering knowledge is necessary to set up the model input specifications in some cases. We also emphasize that these answers are different from what would be obtained using standard EPA plume rise and Gaussian dispersion equations for neutrally or positively buoyant stack gas emissions.

3.1 Scenario 1: Elevated Release of Normal Butane

This scenario describes the release of normal butane (nC_4) at a rate of 15 kg/s from a 0.75-m (i.d.) vent stack at 5-m elevation (Figure 3-1). The vent stack points directly upward and it is assumed that there is no interference from nearby buildings. The distance to the point at which the dense plume sinks to the ground and the concentration at that point are of interest, as well as the distance to the lower flammable limit (LFL) and 25% LFL ground level concentration. Instantaneous concentrations are assumed to be given by a 10 s averaging time.

Calculations are to be made for surface roughnesses, z_0, of 0.03 and 0.3 m, for stability class D (with wind speeds 2 and 5 m/s), and for stability class F (with wind speed 2 m/s). These wind speeds are assumed to apply at stack height. These conditions are chosen to provide upper and lower bounds on solutions for typical roughnesses and stabilities found in routine applications.

The Hoot/Meroney/Peterka (HMP) model is to be used to calculate the trajectory and concentrations in the initial plume, and the Gaussian, SLAB, and DEGADIS models are to be used to calculate transport and dispersion after the initial plume touches the ground.

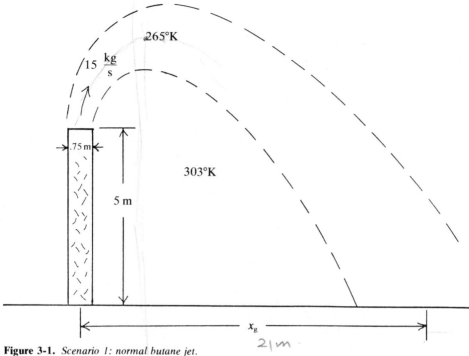

Figure 3-1. *Scenario 1: normal butane jet.*

3.1.1 *Source Emissions*

The mass emission rate and the jet diameter given for this scenario are sufficient to estimate the plume velocity. No further calculations (flashing, pool evaporation, etc.) are required to define the emissions in Scenario 1. Other given parameters are listed below. We note that gas densities can be converted from one temperature to another through the formula $\rho_1/\rho_2 = T_2/T_1$, where temperature T must be in absolute units (e.g., °K).

Material: Normal Butane at Ambient Pressure and 265°K
    Molecular weight $M_g = 58.12$ kg/kg-mol
    Density of gas $\rho_g = 2.67$ kg/m³ at 1 atm and 265°K
    Specific heat $c_p = 1715$ J/kg/°K

Site Data: $z_0 = 0.03$ m; 0.3 m

Ambient Meteorology:
    Temperature $T_a = 303$°K
    Pressure $p_a = 1.013 \times 10^5$ N/m²
    Relative humidity   = 50%
    Class D $u = 2$ m/s ⎫
    Class D $u = 5$ m/s ⎬ at stack height
    Class F $u = 2$ m/s ⎭
    Density $\rho_a = 1.16$ kg/m³

Stack Data:
    Stack height
    Inner diameter
    Emission rate
    Exit velocity

$h_s = 5$ m
$D = 0.75$ m
$Q = 15$ kg/s
$w_0 =$ Volume flow rate/stack area
$= [(15 \text{ kg/s})/(2.67 \text{ kg/m}^3)]/[\pi(0.75 \text{ m}/2)^2]$
$= 12.7$ m/s

To decide whether a dense-gas model is really necessary in this scenario, it is necessary to calculate the source Richardson number, Ri_0, and determine whether it exceeds 10. Havens and Spicer (1985) define $Ri_0 = [g(\rho_g - \rho_a)/\rho_a](\pi/4)Dw_0/uu_*^2$ for continuous sources. If it is assumed that the limiting wind speed is 5 m/s and that $(u_*/u) = 0.065$ (Hanna *et al.*, 1982) then $Ri_0 = 180$, thus justifying the use of dense-gas models.

3.1.2 *Application of HMP Model for Near-Field Transport and Dispersion*

This scenario is ideally suited for application of the Hoot/Meroney/Peterka (HMP) model, which was developed in 1973 by those three authors from wind tunnel observations. This model consists of a small set of analytical equations describing the trajectory and concentration impacts of a dense-gas jet emitted vertically into a crosswind.

The following equations are used to calculate the maximum rise above the stack, Δh, the downwind distance to plume touchdown, x_g, and the concentration ratio C/C_0 at plume touchdown [these same equations were given earlier as Eqs. (2-16) through (2-18)]:

$$\Delta h/2R_0 = 1.32(w_0/u)^{1/3}(\rho_0/\rho_a)^{1/3}\{w_0^2\rho_0/[2R_0g(\rho_0 - \rho_a)]\}^{1/3} \qquad (3\text{-}1)$$

$$\begin{aligned} x_g/2R_0 = &\{w_0u\rho_0/[2R_0g(\rho_0 - \rho_a)]\} \\ &+ 0.56\{(\Delta h/2R_0)^3[(2 + h_s/\Delta h)^3 - 1]\}^{1/2}\{u^3\rho_a[2R_0gw_0(\rho_0 - \rho_a)]\}^{1/2} \end{aligned}$$
$$(3\text{-}2)$$

$$C/C_0 = 2.43(w_0/u)[(h_s + 2\Delta h)/2R_0]^{-1.95} \qquad \text{(at the point } x_g) \qquad (3\text{-}3)$$

where w_0, ρ_0, and R_0 are the initial plume speed, density, and radius, respectively. The variables u and ρ_a are the ambient wind speed and density, respectively.

The initial gas density, ρ_0, in our example is 2.67 kg/m³ at 265°K, and the air density, ρ_a, is 1.16 kg/m³ at 303°K. The initial radius, R_0, is 0.375 m, and the initial velocity, w_0, is 12.7 m/s. Application of Eqs. (3-1)–(3-3) leads to the results:

	$u = 2$ m/s	$u = 5$ m/s
x_g	21 m	57 m
C/C_0 at x_g	0.02244	0.01393
C at x_g	0.0525 kg/m³	0.0326 kg/m³

The ratio C/C_0 can be considered equivalent to either the mole fraction or mass fraction of normal butane at touchdown. It is assumed that the temperature and pressure of the plume are unchanged during its trajectory. Hence, the corresponding concentrations expressed as percentage by volume are just 2.24 and 1.39% for wind speeds of 2 and 5 m/s, respectively. The mass concentrations listed above are computed for the ambient temperature of 303°K, and so are obtained from the product of the mole fraction and the density of normal butane at 303°K (2.338 kg/m³).

The LFL for normal butane is about 1.90% volume. Therefore the HMP model predicts that the normal butane concentration at plume touchdown is about 1.18 times the LFL for $u = 2$ m/s and 0.73 times the LFL for $u = 5$ m/s.

3.1.3 *Far-Field Transport and Dispersion Calculated by Gaussian, SLAB, and DEGADIS Models*

The three models that are applied to calculate transport and dispersion after plume touchdown are the Gaussian model (for nondense gases) and the SLAB and DEGADIS models (for dense gases).

GAUSSIAN MODEL

The concentration, C_{max}, predicted by the HMP model at touchdown (x_g) can be matched with the concentration predicted by the Gaussian model $C = Q/(\pi u \sigma_{y0} \sigma_{z0})$ to give

$$\sigma_{y0} = \sigma_{z0} = [Q/(\pi u C_{max})]^{1/2} \tag{3-4}$$

This relationship is obtained by (1) conserving the total mass flux, $\int_0^\infty \int_{-\infty}^\infty uC \, dy \, dz$, (2) assuming that full Gaussian plume reflection is occurring, and (3) assuming that the lateral and vertical dispersion parameters are equal. If the assumption of full reflection were modified and a circular plume cross section were assumed instead, the right-hand side of Eq. (3-4) would be divided by $2^{1/2}$. Equation (3-4) yields $\sigma_{y0} = \sigma_{z0} = 6.74$ m for $u = 2$ m/s and $\sigma_{y0} = \sigma_{z0} = 5.41$ m for $u = 5$ m/s. The effective averaging time for this problem is assumed to be 10 s or 0.17 min, yielding a correction factor for σ_y of 0.44 [from Eq. (2-21)]. Solving for the virtual source distances, x_{vy} and x_{vz}, using the equations in Section 2.4, we find the following values:

	$u = 2$ m/s		$u = 5$ m/s
	Class D	*Class F*	*Class D*
σ_y virtual source	$x_{vy} = 193$ m	$x_{vy} = 390$ m	$x_{vy} = 155$ m
σ_z virtual source	$x_{vz} = 122$ m	$x_{vz} = 482$ m	$x_{vz} = 96$ m

As an illustration of the calculation of the virtual source distance, x_{vy}, consider the case in which the wind speed, u, equals 5 m/s and the stability class is D. From Eq. (3-4), it is seen that $\sigma_{y0} = 5.41$ m. Equation (2-22) should be used to calculate x_{vy}, with $a = (0.08)(0.17 \text{ min}/10 \text{ min})^{0.2} = (0.08)(0.44) = 0.035$, where the 0.08 parameter comes from the rural class D σ_y equation for averaging times of 10 min in Table 2-2, $\sigma_y = 0.08x/(1 + 0.0001x)^{1/2}$. This equation also shows that $b = 0.0001$. Then Eq. (2-22) can be solved:

$$x_{vy} = (b\sigma_{y0}^2 + \sqrt{b^2\sigma_{y0}^4 + 4a^2\sigma_{y0}^2})/2a^2$$
$$= [10^{-4}(5.41)^2 + \sqrt{10^{-8}(5.41)^4 + 4(0.035)^2(5.41)^2}]/2(0.035)^2$$
$$= 155 \text{ m}$$

These virtual source distances are all greater than the actual touchdown distance, x_g, since the Gaussian plume spread rate is less than the jet spread rate. These distances are then used to estimate σ_y and σ_z at various actual downwind distances from the source. The Gaussian equation is applied to calculate the ground level maximum concentration at several downwind distances. The locations of the LFL ($C = 0.0444 \text{ kg/m}^3$ or 1.90% volume) and the 1/4 LFL ($C = 0.0111 \text{ kg/m}^3$ or 0.48% volume) are then found by interpolation. The results are reported in Table 3-1 along with the results from SLAB and DEGADIS, which will be discussed later. We emphasize that an averaging time of 10 s is used in these calculations. The net effect of this reduced averaging time is a reduction in σ_y, or an increase in concentration, at any distance by a factor of 2.27 over that appropriate for 10 min averages.

TABLE 3-1
Results of Applying a Gaussian Plume Model (GPM) and the SLAB and DEGADIS (DEG) Models to Scenario 1 (Normal Butane)

z_0 (m)	Stability class	Wind speed (m/s)	Touchdown distance (m)	Distance to LFL (m)			Distance to 1/4 LFL (m)		
				GPM	SLAB	DEG	GPM	SLAB	DEG
0.03	F	2	21	62	57	62	614	320	236
0.3	F	2	21	62	44	63	614	215	149
0.03	D	2	21	35	54	78	221	305	242
0.3	D	2	21	35	44	64	221	185	144
0.03	D	5	57	$C <$ LFL			149	145	135
0.3	D	5	57	at touchdown			149	95	92

SLAB MODEL

It is more difficult to match the SLAB model than the Gaussian model at the transition point, x_g, of the HMP model, since the SLAB model requires a ground level area source as an initial condition. However, when dense-gas effects are important, the SLAB model is more scientifically realistic than the Gaussian model. In this application, we assume that the vertically oriented cross-sectional area of the HMP plume at touchdown, x_g, is equal to the horizontally oriented area required as input to the SLAB model. The cross-sectional area, A_e, of the HMP plume can be approximated from

$$A_e = Q/(uC_{\max}) \tag{3-5}$$

Consequently $A_e = 143\ \mathrm{m}^2$ for $u = 2\ \mathrm{m/s}$ and $A_e = 92\ \mathrm{m}^2$ for $u = 5\ \mathrm{m/s}$.

These areas are input to the SLAB model as the required "pool area." Then the SLAB model is used to calculate maximum ground level concentrations as a function of downwind distance, x_{SLAB}. At some downwind distance, x_{SO}, the concentration C_{SLAB} matches the HMP maximum concentration, C_{max}, at touchdown, x_g. The SLAB concentrations are thereafter assumed to apply at an actual downwind distance $x = x_{\mathrm{SLAB}} - (x_{\mathrm{SO}} - x_g)$.

The data required to run SLAB are listed below. The variable names correspond to those used in the computer code. Note that the values are those for simulating the case for stability class F and surface roughness equal to 0.3 m (see Section 2 of the Appendix for a listing of the output for this case).

.05812	wms	molecular weight of source gas (kg/g mole)
1715.	cps	heat capacity at constant p (j/kg-°K)
303.0	ts	temperature of source gas (°K)
15.0	qs	mass source rate (kg/s)
143.	as	source area (m²)
10.	avt	averaging time (sec)
1000.	xffm	maximum downwind distance (m)
1.0	zp(2)	concentration measurement height (m)
.0	zp(3)	concentration measurement height (m)
.0	zp(4)	concentration measurement height (m)
0.3	z0	surface roughness height (m)
10.0	za	ambient measurement height (m)
2.0	ua	ambient wind speed (m/s)
303.	ta	ambient temperature (°K)
0.04733	ala	inverse Monin-Obukhov length (1/m)
-1.	terminate	

The inverse Monin–Obukhov length $(1/L)$ in the above data list is a stability parameter that ranges from about -0.1 for very unstable conditions, to 0.0 for neutral conditions, to about $+0.1$ for very stable conditions. A meteorological "preprocessor" program called STABUT is provided by the authors of SLAB for computing the inverse Monin–Obukhov length, $1/L$. The output from this program gives

$$\begin{aligned}
&\mathrm{D(all):} &&1/L = 0 \\
&\mathrm{F}(u = 2\ \mathrm{m/s},\ z_0 = 0.03\ \mathrm{m}): &&1/L = 0.07025\ \mathrm{m}^{-1} \\
&\mathrm{F}(u = 2\ \mathrm{m/s},\ z_0 = 0.30\ \mathrm{m}): &&1/L = 0.04733\ \mathrm{m}^{-1}
\end{aligned}$$

Values for the inverse Monin–Obukhov length can also be obtained as a function of stability and roughness from nomograms published by Golder (1972). The SLAB model produces maximum ground level concentrations as a function of downwind distance. Distances to the LFL and 1/4 LFL are reported in Table 3-1. These distances are within $\pm 30\%$ of the distances calculated by the Gaussian model for all stability class D combinations and for the stability class F example for LFL. But the Gaussian model distance to the 1/4 LFL for the stability class F case is twice the SLAB model distance. These differences can be more or less, or can change sign depending on the release scenario. Next the DEGADIS model results will be given.

DEGADIS MODEL

Like the SLAB model, the DEGADIS model is initialized with a ground level area source. In this scenario the model must be initialized using the HMP model jet predictions at the location of plume touchdown. The initialization method used with SLAB is also followed here. Let A_e be the effective size of the area source, and assume that the plume is traveling at the speed of the air flow (either 2 or 5 m/s). Then the cross-sectional area A_e is calculated from Eq. (3-5),

$$u = 2 \text{ m/s:} \quad A_e = 143 \text{ m}^2 \text{ and radius } R = 6.74 \text{ m}$$

$$u = 5 \text{ m/s:} \quad A_e = 92 \text{ m}^2 \text{ and radius } R = 5.41 \text{ m}$$

An example of the output from DEGADIS is given in Section A1 of the Appendix.

DEGADIS versions 1.5 and 2.0 have an option for specifying averaging time. We are using version 1.4 here, which has no provisions for averaging time variations. However, the averaging time can be accounted for by modifying the formula for lateral dispersion, σ_y, in DEGADIS 1.4, following procedures recommended by the DEGADIS model authors. For steady-state simulations, all DEGADIS model versions use

$$\sigma_y = \delta x^\beta \tag{3-6}$$

The value of δ is prescribed within the DEGADIS 1.4 model as a function of the stability class, and can be assumed to apply to averaging times of 10 min. For averaging times, t_a, other than 10 min, δ should be altered as follows:

$$\delta_{t_a} = \delta_{10 \text{ min}} \left(\frac{t_a}{10 \text{ min}} \right)^{0.2} \tag{3-7}$$

With this correction, the solution is nearly equivalent to that given by DEGADIS 1.5. This can be done during the input phase of executing DEGADIS when the user is given a chance to modify the constants provided by the model in response to the meteorological inputs.

For 10 s (instantaneous) averaging time

$$\delta_{10 \text{ s}} = \delta_{10 \text{ min}} \left(\frac{10 \text{ s}}{600 \text{ s}} \right)^{0.2} = \delta_{10 \text{ min}} 0.441 = 0.0282 \quad \text{(F stability)}$$

$$= 0.0573 \quad \text{(D stability)}$$

Because the temperature of the plume is virtually equal to the ambient temperature, DEGADIS is run in its isothermal mode. The data required by the model when using this option are summarized in the standard listing produced by the model. A sample listing for one of the runs from this scenario is reproduced in the Appendix.

DEGADIS was run for six combinations of wind speed, stability class, and roughness length. The results of the DEGADIS runs are tabulated in Table 3-1.

Actual downwind distances were calculated from

$$x \text{ (actual)} = x \text{ (DEGADIS)} - (x_{DO} - x_g) \tag{3-8}$$

where x_g is the HMP plume touchdown distance and x_{DO} is the distance at which the DEGADIS model matches the HMP concentration. The distances obtained from the DEGADIS simulations are similar to those for the SLAB and Gaussian models. The SLAB and DEGADIS models agree within about ±20% for the 2 m/s winds and are within about ±7% for the 5 m/s winds. The simple Gaussian model predictions are in fair agreement with the SLAB and DEGADIS model predictions because, in this scenario, the influence of the density difference is quickly lost in the rapidly expanding momentum jet.

NOTES

NOTES

NOTES

3. Application of Models to Five Scenarios

3.2 Scenario 2: Pressurized Liquid Release of Anhydrous Ammonia

This scenario describes the ground level release of anhydrous ammonia from (1) a nozzle failure and (2) a line rupture near the bottom of a 40-mT storage tank at ambient temperature (298°K) and 8.9 bar(a) (i.e., at saturation pressure). The 1.5-in.-i.d. nozzle is very near the tank and the line rupture is on a 1.5-in.-i.d. line at a distance of 150 in. from the tank. These two cases illustrate the extremes of all liquid emissions (from the nozzle) and fully developed two-phase flow (from the line rupture). Figure 3-2 illustrates this scenario.

Calculations of maximum ground level concentrations for 15 min and 30 min averaging times are desired, as a function of downwind distance.

The Bernoulli and Fauske and Epstein (1987) equations are to be used for the source term calculations, and the Gaussian and DEGADIS models for downwind dispersion.

3.2.1 *Source Emissions*

The Bernoulli and Fauske and Epstein (1987) equations can be used to calculate mass emission rate, but require the following input parameters:

Material: Anhydrous Ammonia (NH_3)
  Latent heat of vaporization $\Lambda = 1.17 \times 10^6$ J/kg   (Fauske and Epstein, 1987)
  Specific heat at constant
    pressure for liquid $c_{pl} = 4.49 \times 10^3$
    ammonia   J/kg/°K   (Fauske and Epstein, 1987)
  Density of gas at
    $p = 8.9$ bar and $\rho_g = 6.2$ kg/m³
    $T = 298$°K

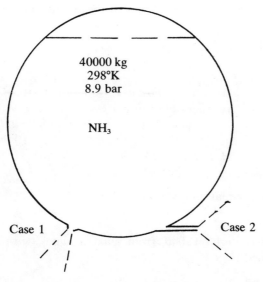

Figure 3-2. *Scenario 2: anhydrous ammonia release. Case 1 is a nozzle failure and Case 2 is a line rupture. In both cases, the inner diameter of the opening is 1.5 in.*

Density of liquid at
  $p = 8.9$ bar and $\rho_l = 608.76$ kg/m^3
  $T = 298°K$
Normal boiling point $T_b = 240°K$ (Perry and Green, 1984)

Tank Conditions: 40 mT = 40,000 kg
  Mass in tank $T_s = 25°C$ (298°K)
  Temperature $p_s = 8.9$ bar (8.9×10^5 N/m^2) (saturation)
  Pressure

Site Conditions: $z_0 = 0.03$ m; 0.30 m

Ambient Meteorology:
  Temperature $T_a = 25°C$ (298°K)
  Pressure $p_a = 1.013 \times 10^5$ N/m^2 (= 1 atm = 1.013 bar)
  Relative humidity   = 80%
  Specific heat at constant
    pressure $c_p = 1004$ J/kg/°K
  Density
    $p_a = 1$ atm and $T_a = 298°K$ $\rho_a = 1.184$ kg/m^3
  Class D $u = 2$ m/s ⎫
  Class D $u = 5$ m/s ⎬ at release height
  Class F $u = 2$ m/s ⎭

Rupture Data:
  Nozzle and pipe
    diameter $D = 1.5$ in. $= 3.81 \times 10^{-2}$ m
  Pipe length $L_p = 150$ in. $= 3.81$ m

These input parameters are used in the following equations:

Case 1. Nozzle at tank assumes all flow at the nozzle aperture is liquid [see Eq. (2-3)].

Bernoulli: $Q_l = C_D A \rho_l \left(2 \dfrac{\Delta p}{\rho_l} + 2gH \right)^{1/2}$

where $C_D =$ $= 0.6$
         $A =$ area of nozzle $= 1.14 \times 10^{-3}$ m^2
         $\rho_l =$ density of liquid
              in tank $= 608.76$ kg/m^3
         $\Delta p = p_{tank} - p_{atm}$ $= 7.9 \times 10^5$ N/m^2
         $H =$ height of liquid $= 4$ m
         $g = 9.8$ m/s^2

Bernoulli's equation gives $Q_l = 21.5$ kg/s.

Case 2. Pipeline ruptured 150 in. from tank (assumes two-phase flow has been established).

Two-phase
flow: $Q_t = F \left(\dfrac{A\Lambda}{1/\rho_g - 1/\rho_l} \right) (T_s c_{pl})^{-1/2}$ [see Eq. (2-4)]

assume Λ = latent heat of
                          vaporization of NH_3 = 1.17×10^6 J/kg
                    ρ_g = gas density in tank   = 6.2 kg/m^3
                    ρ_l = liquid density of NH_3 = 608.76 kg/m^3
                    T_s = tank storage temperature = 298°K
                    c_{pl} = liquid specific heat   = 4.49×10^3 J/kg/°K
                      F = 0.75 (from Table 2-1)
                      $A = \pi D^2/4 = 1.14 \times 10^{-3}$ m^2

This equation gives Q_t = 5.42 kg/s, which is 25% of the emission rate in Case 1 given by Bernoulli's equation for an all-liquid emission.

The time to empty the tank of its liquid contents, assuming the liquid properties are unchanged, is 31 min for the nozzle at the tank and 123 min for the pipeline ruptured 150 in. from the tank. The mass fraction that is flashed, $f = c_{pl} (\Delta T/\Lambda)$, equals 0.223, where $\Delta T = (T_s - T_b)$ equals 298 − 240 or 58°K. If all unflashed liquid is entrained into the gas jet as an aerosol, and the cloud temperature is T_b = 240°K, the effective density of the aerosol cloud is given by the relation

$$\bar{\rho} = \frac{1}{(1 - f)/\rho_l + f/\rho_g(240°K)} = 3.86 \text{ kg/m}^3 \qquad (3\text{-}9)$$

This is the initial density prior to any air entrainment.

To decide whether or not dense-gas effects are important in this scenario, the source Richardson number, Ri_0, can be calculated. Defining $Ri_0 = g[(\rho_g - \rho_a)/\rho_a](\pi/4)Dw_0/uu_*^2$ for continuous sources (Havens and Spicer, 1985), and assuming that the limiting wind speed is 5 m/s and $(u_*/u) = 0.065$, then $Ri_0 = 43$, which exceeds the criterion ($Ri_0 \cong 10$) for including dense-gas effects. To obtain this result it is further assumed that about 20% of the liquid flashes, giving a source emission rate of about 4 kg/s for the gas, and that the effective diameter, D, after expansion is about 1 m. Consequently, the initial plume velocity, w_0, after expansion would be about 1.3 m/s.

3.2.2 Applications of Gaussian and DEGADIS Models to Calculate Downwind Transport and Dispersion

GAUSSIAN MODEL
The simple Gaussian model does not consider the thermodynamic effects but treats the release as an inert, nonbuoyant gas with continuous source strength of 21.5 kg/s in Case 1 and 5.42 kg/s in Case 2. Concentrations for 15 and 30 min averaging times are calculated using the correction $\sigma_y(t_a) = \sigma_y$ (10 min) $\times$ (10 min/$t_a)^{-0.2}$, as outlined in Chapter 2. The resulting predicted Gaussian concentrations are listed in Tables 3-2 and 3-3 for the two cases in Scenario 2, for representative downwind distances ranging from 0.1 to 2.0 km.

In the case of the Gaussian model, the ratios of concentrations for Case 1 (nozzle at tank) to those for Case 2 (pipe rupture) are proportional to the ratio of the source terms. The ratio of concentrations for 30 and 15 min averaging times in the tables is also a constant (15 min/30 min)$^{0.2}$ = 0.87. Concentrations decrease

TABLE 3-2
Gaussian Model Concentration Predictions in ppm for Scenario 2, Case 1
(Nozzle at Tank)

Downwind distance (km)	15 min averaging time			30 min averaging time		
	Class F 2 m/s	Class D 2 m/s	Class D 5 m/s	Class F 2 m/s	Class D 2 m/s	Class D 5 m/s
0.1	732,000	102,000	40,600	696,000	96,700	38,700
0.2	207,000	29,600	11,900	180,000	25,900	10,300
0.5	36,400	5,580	2,240	31,700	4,860	1,940
1.0	10,500	1,710	684	9,170	1,490	596
1.5	5,350	886	354	4,660	771	309
2.0	3,390	564	226	2,950	491	196

TABLE 3-3
Gaussian Model Concentration Predictions in ppm for Scenario 2, Case 2 (Pipe Rupture)

Downwind distance (km)	15 min averaging time			30 min averaging time		
	Class F 2 m/s	Class D 2 m/s	Class D 5 m/s	Class F 2 m/s	Class D 2 m/s	Class D 5 m/s
0.1	185,000	25,700	10,200	175,000	24,400	9,760
0.2	52,200	7,460	3,000	45,400	6,530	2,600
0.5	9,180	1,410	565	7,990	1,230	489
1.0	2,650	431	172	2,310	376	150
1.5	1,350	223	89.2	1,170	194	77.9
2.0	855	142	57.0	744	124	49.4

by a factor of about six as stability class changes from F to D, and decrease by
a factor of about 2.5 as wind speed changes from 2 to 5 m/s. The concentration
predictions are unrealistically high (close to 10^6 ppm or 100%) close to the
source because the diluting influence of the initial jet is neglected.

These Gaussian model predictions will be compared with the DEGADIS
model predictions in the next subsection.

DEGADIS MODEL

The DEGADIS model is intended for application to area source emission of a
gas. Versions 1.4 and 1.5 of the model do not explicitly treat either jets or
entrained liquids. A later version (2.0) of DEGADIS treats jets, but was not
officially released at the time these calculations were being made. In this scenario
the source is a jet at its normal boiling point of 240°K, with 22.3% of the jet mass
as gas and 77.7% of the jet mass as liquid. Observations from the Desert Tortoise
Test show that the ammonia liquid from such a release exists as an aerosol that is
carried downwind in the jet and does not "rain out" on the surface (Goldwire *et
al.*, 1985).

The density of the two phase jet at the source was calculated in Section
3.2.1 to be 3.86 kg/m^3 at a temperature of 240°K. For the purposes of initializing
the DEGADIS model, it is assumed that the initial jet is a "pseudo-gas" that is all

gas at ambient temperature of 298°K. Equation (3-9) is used once again to compute the effective density of this "pseudo-gas," but now the density of the gas (ρ_g) is taken at 298°K rather than 240°K, so that $\rho_e = 3.11$ kg/m^3. Note that the volume of the "pseudo-gas" is controlled by the volume of gas-phase ammonia, so that the effective density at 298°K is virtually identical to $(3.86 \text{ kg/m}^3) \times (240°K/298°K)$. The effective molecular weight M_e of the "pseudo-gas" is

$$M_e = \rho_e \times \text{molar volume}$$

$$= (3.11 \text{ kg/m}^3)(22.4 \text{ m}^3/\text{kg-mol})(298°K/273°K)$$

$$= 76.04 \text{ kg/kg-mol}$$

It is important to note that all the thermodynamic implications of this assumption are being neglected, the model is run in isothermal mode, and the density profile has been approximated. We also note that alternate methods for starting the DEGADIS model are possible. One could add sufficient air adiabatically to evaporate all the ammonia aerosol. Another method would be to use the DEGADIS option of providing a density and concentration profile. In this case a number of computations would be required in which various quantities of air are mixed into the ammonia aerosol.

The effects of averaging time, t_a, are accounted for by specifying a revised δ in the DEGADIS model input stream [revised $\delta = \delta$ (10 min) $\times (t_a/10 \text{ min})^{0.2}$]. This δ is the coefficient in the σ_y formula:

	δ (15 min)	δ (30 min)
Class D	0.141	0.162
Class F	0.0694	0.0797

The source is treated as a circular area source with radius arbitrarily selected to be 1 m. Therefore, the calculations are not expected to be very accurate within about 100 m of the source location, where the effects of the source parameters are important. The model generates a shallow "source blanket" of high-concentration gas over the source with a radius of over 200 m for the Class F case (see the example listing in Section 3 of the Appendix). Model runs were carried out for 24 independent cases:

2 source strengths (Case 1 and Case 2)
2 averaging times (15 and 30 min)
2 roughness lengths (0.03 and 0.3 m)
3 wind-speed/stability combinations.

It would take too much space to present tables or graphs of downwind concentration distributions for each of the 24 cases. Instead, the predicted concentration for the 24 DEGADIS model runs at a downwind distance of 1.0 km is listed in Table 3-4, along with the corresponding prediction from the Gaussian model.

Note that these concentrations are expressed in units of g/m^3 rather than ppm by volume. This is because the best measure of the amount of ammonia in

TABLE 3-4
Predicted Concentrations (g/m³) at a Distance 1 km Downwind in Scenario 2 (Ammonia)

	15 min averaging time			*30 min averaging time*		
	DEGADIS			*DEGADIS*		
	$z_0 = 0.03\ m$	$z_0 = 0.3\ m$	*Gaussian*	$z_0 = 0.03\ m$	$z_0 = 0.3\ m$	*Gaussian*
	Case 1 ($Q = 21.5$ kg/s)					
Class F, 2 m/s	3.50	1.22	6.72	3.50	1.22	5.85
Class D, 2 m/s	3.02	1.17	1.09	3.02	1.13	0.949
Class D, 5 m/s	0.933	0.443	0.436	0.836	0.390	0.380
	Case 2 ($Q = 5.42$ kg/s)					
Class F, 2 m/s	1.08	0.434	1.69	1.08	0.440	1.47
Class D, 2 m/s	0.778	0.301	0.275	0.691	0.272	0.239
Class D, 5 m/s	0.220	0.113	0.110	0.191	0.092	0.096

the air is its mass, rather than the number of parts of the vapor/aerosol "pseudo-gas" per million parts of the vapor/aerosol/air mixture.

Table 3-4 shows that the Gaussian and DEGADIS ($z_0 = 0.3$ m) predictions are within about 10% of each other for the stability class D tests at the specified downwind distance of 1 km. The DEGADIS model shows a decrease in concentration of a factor of about two to three as roughness length increases from 0.03 to 0.3 m. For the stable class F example, the Gaussian model predictions are

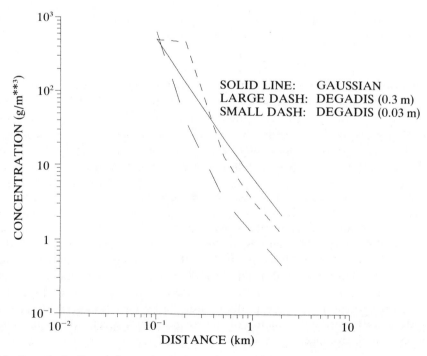

Figure 3-3. *Scenario 2, Case 1 (ammonia discharge from nozzle). Predictions of DEGADIS and Gaussian models for class F, u = 2 m/s, and averaging time of 15 min.*

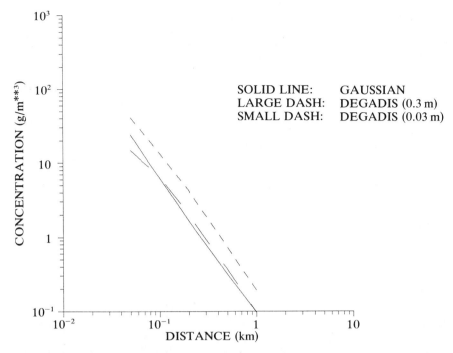

Figure 3-4. *Scenario 2, Case 2 (ammonia discharge from pipe rupture). Predictions of DEGADIS and Gaussian models for class D, u = 5 m/s, and averaging time of 30 min.*

higher than the DEGADIS model predictions by a factor of two to four, presumably due to the influence of the source blanket in the DEGADIS model. The source blanket has a maximum half-width of 209 m in that case.

Figures 3-3 and 3-4 show the variation of predicted concentration with downwind distance for some of the cases that were run. During stable conditions, when there is an extensive source blanket predicted by the DEGADIS model, as in Figure 3-3, the Gaussian and DEGADIS curves show markedly different behavior, as the DEGADIS predictions "dive" from a 100% gas concentration at the edge of the source blanket. In this range the DEGADIS predictions decrease by two orders of magnitude as distance increases by only a factor of two.

During neutral conditions, when there is a very small source blanket predicted by the DEGADIS model, as in Figure 3-4, the Gaussian and DEGADIS curves have similar slopes and magnitudes. In fact the Gaussian and DEGADIS ($z_0 = 0.3$ m) curves are within ±20% of each other at all distances plotted on this figure. However, this good agreement is true only if a roughness length of 0.3 m is assumed in DEGADIS. Most of the experiments on which the Gaussian curves are based were conducted over surfaces with roughnesses of about 0.1 to 0.3 m. As seen in the figure, the DEGADIS curve for $z_0 = 0.3$ m is shifted downward by a factor of approximately two relative to the curve for 0.03 m at all distances.

NOTES

NOTES

NOTES

3.3 Scenario 3: High-Pressure Gas Release of Carbon Monoxide

This scenario describes the release of high-pressure (2230 psig) CO from a horizontally directed valve with a valve seat diameter of 1.27 cm and a discharge pipe diameter (initial jet diameter) of 10.2 cm (see Figure 3-5). This combination of valve seat diameter and discharge pipe diameter will ensure that the pipe exit velocity is less than or equal to the sonic velocity. The duration of the release is assumed to be 5 min. Maximum ground level centerline concentrations as a function of downwind distance must be calculated for 1 and 15 min averaging times. The initial direction of the jet is assumed to be perpendicular to the wind direction.

The critical flow equation is to be used to calculate the source emission rate, the Ooms model is to be used for analysis of the jet, and the Gaussian model used for subsequent downwind transport and dispersion.

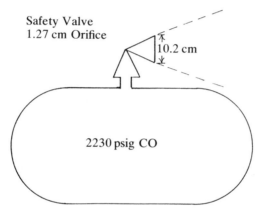

Figure 3-5. *Scenario 3: carbon monoxide jet, with initial sonic velocity.*

3.3.1 *Source Emissions*

The following input parameters are needed to calculate the source emission rate:

Properties of Carbon Monoxide:

Molecular weight (CO) $\qquad M_g = 28.0$ kg/kg-mol

CO density $\qquad \rho_s = 1.30$ kg/m^3 ($T_g = 262°$K, $p = 1$ atm)

[Assumes isenthalpic depressurization, neglecting kinetic energy effects, as CO starts at $p = 154$ atm and 294°K and ends at $p = 1$ atm and 262°K, as interpolated by eye on Figure 3-20, p. 3-171, in Perry and Green (1984). There are alternative approaches such as assumption of isentropic behavior, in which kinetic energy effects are included.]

Tank Data:

Temperature $\qquad T_s = 294°$K

Pressure $\qquad p_s = 1.54 \times 10^7$ N/m^2

Valve seat diameter $\qquad = 0.0127$ m

Discharge pipe diameter $\quad D = 0.102$ m

Site Data: $z_0 = 0.03$ and 0.3 m

Air Density: $\rho_a = 1.20$ kg/m^3

Standard equations for gas flow from orifices are used to calculate the mass flux. The criterion for critical flow is

$$p/p_a \geq [(\gamma + 1)/2]^{\gamma/(\gamma-1)}$$

[Hanna and Drivas, 1987, Eq. (4-1), and Eq. (2-8) of this document] where p and p_a are tank and ambient pressures and γ is c_p/c_v (1.4 for CO; Perry and Green, 1984, p. 3-144). For this case $p/p_a = 154$, which is much greater than 1.9, and hence critical flow exists.

The emission rate with critical flow [Hanna and Drivas, 1987, Eq. (4-2), and Eq. (2-9) of this document] is calculated in the following manner:

$$Q = 0.8 Ap\{(\gamma M/R^* T_s)[2/(\gamma + 1)]^{(\gamma+1)/(\gamma-1)}\}^{0.5}$$

where $A = 1.27 \times 10^{-4}$ m^2 (based on valve inner diameter)
$p = 154 \times 10^5$ N/m^2
$R^* = 8314$ J/kg-mol/$^\circ$K
$M = 28$ kg/kg-mol
$T_s = 294^\circ$K

$\therefore$ Source emission rate $Q = 3.63$ kg/s.

The flow then expands to the discharge pipe diameter (0.102 m) and enters the ambient atmosphere with a temperature of 262°K, a density of 1.30 kg/m^3, and a pressure of 1 atm = 10^5 N/m^2. The volume flow rate is given by

$$V = Q/\rho = (3.63 \text{ kg/s})/(1.30 \text{ kg/m}^3) = 2.79 \text{ m}^3/\text{s}$$

and the exit velocity is $w_0 = V/A = (2.79 \text{ m}^3/\text{s})/(0.00817 \text{ m}^2) = 341$ m/s.

The Havens and Spicer (1985) definition of initial Richardson number, $Ri_0 = g[(\rho_g - \rho_a)/\rho_a](\pi/4)Dw_0/uu_*^2$ can be used to determine whether it is important to consider dense-gas effects. Assume that $u = 5$ m/s is the limiting case and further assume that $u_*/u = 0.065$. Then substituting for the other parameters ($\rho_g = 1.30$ kg/m^3, $\rho_a = 1.20$ kg/m^3, $D = 0.102$ m, and $w_0 = 341$ m/s), we obtain $Ri_0 = 42$, which exceeds the critical Ri_0 of 10 and implies that dense-gas effects are indeed important in this scenario.

3.3.2 *Dense Jet Calculations*

Given the source emissions parameters from Section 3.3.1, it is possible to calculate the transport and dispersion of the dense jet. The maximum distance traveled by the plume in a crosswind can be calculated using Briggs's momentum jet formula [Eq. (2-14)]. The exit radius R_0 equals 0.5 times the diameter, or 0.051 m in this scenario. The maximum crosswind distance, $\Delta z = 4.8 w_0 R_0/u$, is calculated to be 42 m for the 2 m/s wind speed and 17 m for the 5 m/s wind speed. Briggs's formulas also show that the downwind distance traveled by the plume when it reaches this maximum crosswind distance is approximately equal to Δz. Therefore we assume that the downwind distance equals Δz for the purpose of this example.

The Ooms dense momentum jet model (as programmed by Havens, 1987) can be used to more precisely simulate the transport and dispersion of the dense jet in the near field. As mentioned above, it is assumed that the crosswind and downwind distances traveled by the jet are equal. The transition from dense jet to Gaussian plume is assumed to occur at the distances calculated by Briggs's formula (42 m for a wind speed of 2 m/s and 17 m for a wind speed of 5 m/s). The Ooms/Havens computer code operates on a PC and is intended to be used as input to the DEGADIS version 2.0 dense-gas dispersion model. Copies of the output from the Ooms/Havens code for the source conditions and two wind speeds in this scenario are given in Table 3-5. Note that the model gives you only the height and concentration on the jet centerline as a function of downwind distance. The plume does not sink more than 0.2 m below its initial height of 2 m in either case, and the centerline concentrations at the transition points are about 0.0161 kg/m^3 for the 2 m/s wind speed and 0.0403 kg/m^3 for the 5 m/s wind speed. The ratio $(\rho_p - \rho_a)/\rho_a$ is therefore about 0.001 and 0.003 at the transition points for these two cases, verifying that dense-gas effects are not important at those downwind distances.

The calculated concentrations at the transition points are used in the next section to initialize the Gaussian plume model.

3.3.3 *Transport and Dispersion Calculated by Gaussian Model*

As discussed in Section 2.4, some calculations need to be made at the jet-plume transition point to initialize the Gaussian plume formula in such a way that the maximum concentration and the mass flux are conserved. From the input conditions and the Ooms/Havens model output, the following facts are known at the transition distance of 42 m for the 2 m/s wind speed and 17 m for the 5 m/s wind speed:

$u = 2$ m/s case	$u = 5$ m/s case
$C_{max} = 0.0161$ kg/m^3	$C_{max} = 0.0403$ kg/m^3
$Q = 3.63$ kg/s	$Q = 3.63$ kg/s

It is assumed that the lateral and vertical dispersion parameters σ_y and σ_z are equal at the transition point. If the height of the plume centerline, h, is right on the ground or if the height of the plume centerline is so far above the ground that the ratio of plume height to vertical dispersion parameter σ_z is much greater than one, then the Gaussian equations relating the dispersion parameters to the source emission rate, the wind speed, and the concentration at the transition point are simplified:

$$\sigma_{y0} = \sigma_{z0} = \sqrt{Q/(\pi u C_{max})} \qquad \text{if } h/\sigma_z \ll 1 \qquad (3\text{-}10)$$

TABLE 3-5
Ooms/Havens Model Output for 5 min Averaging Time for Scenario 3

```
TEST NO.:          3

DESCRIPTION: AIChE    Scenario    3  (u=5)
    O INITIAL DOWNWIND DISTANCE TO FIRST PRINT OUT =        .0000 M
    O INCREMENTAL DOWNWIND DISTANCE BETWEEN PRINTOUTS =       5.0000 M
    O AMBIENT AIR DENSITY =     1.200 KG/M**3
    O DISPERSION JET DENSITY =     1.300 KG/M**3
    O AMBIENT TEMPERATURE = 529.2 DEG RANKINE
    O INITIAL JET DIAMETER =    10.2000 CM
    O INITIAL JET FLOWRATE =    .2790E+01 M**3/S
    O DEPTH OF SECTOR =  1000.000 M
    O DISTANCE FROM THE SECTOR CENTERLINE TO THE JET ORN/8XIN THE Y DIRECTION
=     .000 M
    O DISTANCE FROM THE SECTOR SURFACE TO THE JET ORIGIN
      IN THE Z DIRECTION =    998.000 M
    O LAMBDA COEFFICIENT FOR THE VERY NEAR FIELD MODEL =      1.16190
    O MEAN VELOCITY OF THE WIND =       5.000 M/S
    O ORIENTATION OF THE JET
         HORIZONTAL TRANSVERSE TO THE RIGHT
    O ENTRAIN COEFFICIENTS
         ALFA 1=   .0570
         ALFA 2=   .5000
         ALFA 3= 1.0000

   VERY NEAR FIELD JET CENTERLINE PATH:

        X                Z           CONCENTRATION(CL)
     (METERS)        (METERS)          (KG/M**3)

       .1320           2.000           .1040E+01
      5.0160           2.000           .1238E+00
     10.0320           2.000           .6554E-01
     15.0150           1.998           .4468E-01
     20.0309           1.989           .3377E-01
     25.0137           1.980           .2710E-01
     30.0296           1.970           .2255E-01
     35.0126           1.955           .1927E-01
     40.0288           1.936           .1676E-01
     45.0119           1.918           .1480E-01
     50.0280           1.899           .1322E-01


TEST NO.:          3

DESCRIPTION: AIChE    Scenario    3  (u=2)
    O INITIAL DOWNWIND DISTANCE TO FIRST PRINT OUT =        .0000 M
    O INCREMENTAL DOWNWIND DISTANCE BETWEEN PRINTOUTS =       5.0000 M
    O AMBIENT AIR DENSITY =     1.200 KG/M**3
    O DISPERSION JET DENSITY =     1.300 KG/M**3
    O AMBIENT TEMPERATURE = 529.2 DEG RANKINE
    O INITIAL JET DIAMETER =    10.2000 CM
    O INITIAL JET FLOWRATE =    .2790E+01 M**3/S
    O DEPTH OF SECTOR =  1000.000 M
    O DISTANCE FROM THE SECTOR CENTERLINE TO THE JET ORN/8XIN THE Y DIRECTION
=     .000 M
    O DISTANCE FROM THE SECTOR SURFACE TO THE JET ORIGIN
      IN THE Z DIRECTION =    998.000 M
    O LAMBDA COEFFICIENT FOR THE VERY NEAR FIELD MODEL =      1.16190
    O MEAN VELOCITY OF THE WIND =       2.000 M/S
    O ORIENTATION OF THE JET
         HORIZONTAL TRANSVERSE TO THE RIGHT
    O ENTRAIN COEFFICIENTS
         ALFA 1=   .0570
         ALFA 2=   .5000
         ALFA 3= 1.0000

   VERY NEAR FIELD JET CENTERLINE PATH:

        X                Z           CONCENTRATION(CL)
     (METERS)        (METERS)          (KG/M**3)

       .1320           2.000           .1037E+01
      5.0160           2.000           .1214E+00
     10.0320           2.000           .6405E-01
     15.0150           1.992           .4373E-01
     20.0309           1.981           .3318E-01
```

$$\sigma_{y0} = \sigma_{z0} = \sqrt{Q/(2\pi u C_{max})} \qquad \text{if } h/\sigma_z \gg 1 \qquad (3\text{-}11)$$

The first equation assumes that perfect reflection of the plume occurs at the ground surface. Note that the calculated σ from Eq. (3-10) is 1.41 times that from Eq. (3-11). Initial calculations show that σ_{z0} is on the order of h for the CO jet in Scenario 3, which would require a more complicated solution to be strictly correct, that is, solve for σ_{z0} in the formula

$$C_{max} = [Q/(2\pi u\sigma_{z0}^2)][1 + \exp(-2h^2/\sigma_{z0}^2)] \qquad (3\text{-}12)$$

For the purpose of this example, we assume that complete reflection occurs ($h/\sigma_z \ll 1$) and use Eq. (3-10), which gives $\sigma_{z0} = \sigma_{y0} = 5.99$ m for $u = 2$ m/s and $\sigma_{z0} = \sigma_{y0} = 2.39$ m for $u = 5$ m/s. To match the plume centerline concentrations it must be assumed that the receptor height in the Gaussian calculations equals the plume centerline height in the Ooms model calculations (about 2 m).

With these σ values, the following virtual source distances are calculated using the equations in Section 2.4:

	Virtual distance for		
	σ_y (5 min)	σ_y (1 min)	σ_z
Class D, $u = 2$ m/s	$x_{vy} = 86$ m	$x_{vy} = 119$ m	$x_{vz} = 108$ m
Class D, $u = 5$ m/s	$x_{vy} = 34$ m	$x_{vy} = 47$ m	$x_{vz} = 41$ m
Class F, $u = 2$ m/s	$x_{vy} = 174$ m	$x_{vy} = 240$ m	$x_{vz} = 422$ m

For example, consider the case in which the wind speed, u, equals 5 m/s, the stability class is D, and the virtual distance x_{vz} for $\sigma_{z0} = 2.39$ m must be calculated. In this case, $\sigma_z = 0.06x(1 + 0.0015x)^{-1/2}$ from Table 2-2. Therefore $a = 0.06$ and $b = 0.0015$ and Eq. (2-23) gives

$$x_{vz} = (b\sigma_{z0}^2 + \sqrt{b^2\sigma_{z0}^4 + 4a^2\sigma_{z0}^2})/2a^2 = 41 \text{ m}$$

The two averaging times of interest are 1 and 15 min. But since the duration of the release is only 5 min, the σ_y values have been adjusted to represent 1 and 5 min averages. Because the standard Gaussian model applies to a 10 min averaging time, the adjustment factor for σ_y is either $(10 \text{ min}/5 \text{ min})^{-0.2} = 0.725$ or $(10 \text{ min}/1 \text{ min})^{-0.2} = 0.631$. The concentrations for a 15 min averaging time are obtained by dividing the 5 min concentrations by a factor of 3.0, since the plume exists for only 5 min out of the total 15-min period of interest.

The predicted variation of maximum centerline concentration with down-wind distance is shown in Figures 3-6 and 3-7, for averaging times of 1 and 15 min, respectively. Each figure has separate curves for the three stability class–wind-speed configurations that are studied. The matching procedure at the transition point from the Ooms/Havens model to the Gaussian model produces a fairly smooth curve for all but the class F–wind-speed 2 m/s combination. The knee or plateau in that curve is caused by the virtual source matching procedure,

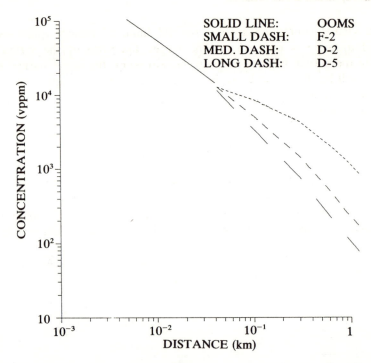

Figure 3-6. *Variation of maximum concentration with distance for a 1 min average in Scenario 3. The Ooms/Havens model is used for small distances, and the Gaussian model for large distances.*

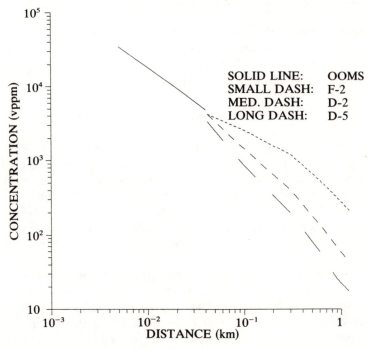

Figure 3-7. *Variation of maximum concentration distance for a 15 min average in Scenario 3. The Ooms/Havens model is used for small distances, and the Gaussian model for large distances.*

since the virtual source distance is several hundred meters while the actual downwind distance is only 45 m. The larger the ratio of the virtual source distance to the actual distance at the transition point, the more marked the knee or plateau will be. In the limit, as the virtual source distance approaches infinity, the slope of the curve approaches zero past the transition point (i.e., it is flat on a diagram such as Figure 3-6).

NOTES

NOTES

NOTES

3.4 Scenario 4: Pressurized Liquid Release of Chlorine

This scenario involves the release of liquid chlorine (Cl_2) from a valve failure 0.3 in. i.d.) on the lower pigtail of a 1-ton cylinder at a temperature of 303°K (see Figure 3-8). The spill occurs on an undiked surface consisting of wet soil. Vapor, aerosol (at various proportions), and a liquid pool must be considered. The maximum ground level concentration for 1 and 15 min averaging times must be calculated as a function of downwind distance.

The source release rate should be calculated using the Bernoulli equation. The SPILLS model should be used to calculate the evaporative flux from the spilled liquid. Subsequent downwind transport and dispersion should be calculated with the CAMEO, SLAB, and DEGADIS models assuming two possible scenarios for aerosol entrainment: Case 1—All liquid that does not flash enters the pool (no aerosol entrainment); Case 2—50% of the liquid that does not flash enters the pool (50% aerosol entrainment). Recent work suggests that there could be as much as 100% aerosol entrainment, but this case is not treated here.

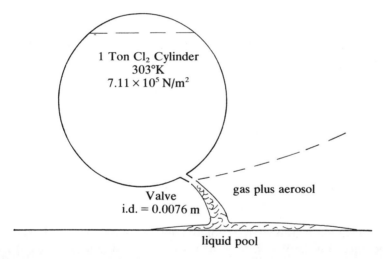

Figure 3-8. *Scenario 4: chlorine spill. The gas phase is made up of flashed vapor and vapor that has evaporated from the liquid pool.*

3.4.1 *Source Emissions*

The following input parameters are needed to calculate the source emission rate.

Properties of Chlorine:

Molecular weight	$M_g = 71.0$ kg/kg-mol
Density of gas	$\rho_g = 2.89$ kg/m^3 at 1 atm and 303°K
Density of liquid	$\rho_l = 1423$ kg/m^3 at 7.11×10^5 N/m^2 and 303°K
Latene heat of vaporization	$\Lambda = 2.88 \times 10^5$ J/kg
Specific heat of the liquid	$c_{pl} = 930$ J/kg/°K at 7.11×10^5 N/m^2 and 239°K
	$= 995$ J/kg/°K at 271°K

Specific heat of the gas $c_p = 480$ J/kg/°K
Normal boiling point $T_b = 239$°K

Liquid Chlorine Storage Data:
  Mass $= 1$ ton or 908 kg (2000 lb)
  Temperature $T_s = 303$°K
  Pressure $p_s = 7.11 \times 10^5$ N/m^2
  Valve inner diameter $D = 0.0076$ m

Ambient Meteorology:
  Temperature $T_a = 303$°K
  Pressure $p_a = 1.013 \times 10^5$ N/m^2
  Relative humidity $= 80\%$
  Class D $u = 2$ m/s
  Class D $u = 5$ m/s
  Class F $u = 2$ m/s

Site Data: $z_0 = 0.03$ m; 0.3 m

The source emission rate, Q_l, is given by Bernoulli's equation (2-1) for all-liquid flow from an opening in a tank.

$$Q_l = C_D A(\rho_l)[2(\Delta p)/(\rho_l) + 2gH]^{0.5}$$

where $C_D = 0.6$
  $A = 4.56 \times 10^{-5}$ m^2
  $\Delta p = 6.1 \times 10^5$ N/m^2
  $\rho_l = 1423$ kg/m^3
  $H = 2$ m

Therefore, $Q_l = 1.17$ kg/s.

It is assumed that the depth of liquid above the hole is 2 m. This assumption has little effect on the flow rate, which is determined mainly by the Δp term. If this flow rate remains steady and it is assumed that the liquid properties do not change with time, then the tank will empty in about 13 min.

Of the total emission rate, Q_l, the percentage flashed is given by $100(c_{pl})(\Delta T)/\Lambda = 21\%$. Therefore an amount $0.79 Q_l$ is spilled as a liquid onto the surface.

The source Richardson number, $Ri_0 = g[\rho_g - \rho_a)/\rho_a] (\pi/4)Dw_0/uu_*^2$ can be calculated to determine whether dense-gas effects are important in Scenario 4. In this scenario, $\rho_g = 2.89$ kg/m^3, and $\rho_a = 1.20$ kg/m^3. The limiting wind speed would be 5 m/s, and it can be assumed that $u_*/u = 0.065$. The mass flux of gas flashed is about 0.25 kg/s, giving a volume flux of about 0.087 m^3/s. It can be arbitrarily assumed that the expansion that takes place during the flashing process results in a plume with a diameter of 0.1 m, implying a plume velocity of 11.1 m/s. Consequently $Ri_0 = 23$, which slightly exceeds the criterion for the consideration of dense-gas effects. This calculation has neglected the contribution of gas evaporated from the liquid pool, which will tend to increase the plume density. Also, Ri_0 will be a factor of 16 larger for the case where the wind speed is 2 m/s.

POOL EVAPORATION

The calculated evaporative emission rates of chlorine from a boiling pool when the wind speed is 2 or 5 m/s suggest a rapid boiling-off of the pool (as modeled by SPILLS). If a pool size is arbitrarily specified, then the resultant emission rate from the pool should be compared with the rate at which *liquid* Cl_2 is entering the pool. If the emission rate from the pool exceeds the rate at which the liquid enters the pool, the initial guess at the size of the pool is too large. By iteration, a steady-state pool area can be arrived at that represents a balance between inflow and evaporation.

Case 1: All liquid that does *not* flash enters the pool (i.e., no aerosol formation).

$$Q(\text{liquid}) = (1 - 0.21)\ 1.17\ \text{kg/s}\ (21\%\ \text{flashes})$$
$$= 0.9243\ \text{kg/s}$$
$$\text{Total mass spilled} = 717.3\ \text{kg (liquid)}$$

After a few iterations, it was determined that the steady-state pool area is 74 m^2 for $u = 2$ m/s and 30 m^2 for $u = 5$ m/s. These areas correspond to radii of 4.8 and 3.1 m, respectively.

Case 2: 50% of the liquid that does *not* flash enters the pool; the rest is entrained as aerosol.

$$Q(\text{liquid}) = 0.5(0.9243\ \text{kg/s}) = 0.4622\ \text{kg/s}$$
$$\text{Total mass spilled} = 348.7\ \text{kg (liquid)}$$

In this case, the iterative approach to determining the steady-state pool area resulted in an area of 38 m^2 for $u = 2$ m/s and an area of 15 m^2 for $u = 5$ m/s. These areas correspond to radii of 3.5 and 2.2 m, respectively.

3.4.2 *Calculation of Transport and Dispersion Using the CAMEO, SLAB, and DEGADIS Models*

CAMEO MODEL

NOAA's CAMEO model contains a standard Gaussian model similar to that used in previous scenarios. It does not account for thermodynamic processes or aerosols, and assumes that the plume has the same temperature and density as its surroundings. It is assumed that the emission rate equals the spill rate (1.17 kg/s) from the cylinder.

CAMEO allows for different qualitative roughness lengths and adjusts the Gaussian model dispersion parameters by a power law relation involving the roughness lengths. In this application, $z_0 = 0.03$ m is assumed to be valid for "thin grass up to 10 cm high," and $z_0 = 0.3$ m is assumed to be valid for "homogeneous forest; suburb full obstacle." CAMEO is valid for averaging times of 10 min since it uses the standard EPA dispersion curves. To convert the CAMEO predictions to averaging times of 1 and 15 min, we made the following

TABLE 3-6
CAMEO Concentration Predictions (ppm) at Various Distances (km) for Scenario 4, Wind Speed 2 m/s, Assuming That the Gas Emission Rate Equals the Spill Rate (1.17 kg/s) from the Cylinder

Concentration (ppm)		Downwind distance (km)			
		Class F, u = 2 m/s		Class D, u = 2 m/s	
1 min averaging	15 min averaging	$z_0 = 0.03\,m$ (km)	$z_0 = 0.3\,m$ (km)	$z_0 = 0.03\,m$ (km)	$z_0 = 0.3\,m$ (km)
3240	1885	0.30	0.15	0.10	0.050
645	379	0.70	0.43	0.25	0.15
324	189	1.10	0.60	0.40	0.22
129	75.4	1.90	1.30	0.62	0.41
64.5	37.9	3.00	2.10	0.90	0.73
12.9	7.54	7.90	6.10	2.40	1.75

corrections [based on Eq. (2-21)]:

$$C(1\,\text{min}) = (10\,\text{min}/1\,\text{min})^{0.2} C(10\,\text{min}) = 1.58\,C(10\,\text{min})$$

$$C(15\,\text{min}) = (10\,\text{min}/15\,\text{min})^{0.2} C(10\,\text{min}) = 0.92\,C(10\,\text{min})$$

It is assumed that the difference is not important between the 15 min averaging time and the 13 min time taken to empty the cylinder.

The CAMEO model calculates "dilution factors," or Cu/Q, which have the units of m^{-2}. Downwind distances at which certain dilution factors are reached are plotted by the model as part of its standard output (see an example in the Appendix). Tables 3-6 and 3-7 give the calculated concentrations using the CAMEO model for the various input conditions for this scenario. As expected, an increase in roughness causes an increase in dilution, such that at large

TABLE 3-7
CAMEO Concentration Predictions (ppm) at Various Distances (km) for Scenario 4, Wind Speed 5 m/s, Class D, Assuming That the Gas Emission Rate Equals the Spill Rate (1.17 kg/s) from the Cylinder

Concentration (ppm)		Downwind distance (km)	
1 min averaging	15 min averaging	$z_0 = 0.03\,m$ (km)	$z_0 = 0.3\,m$ (km)
1297	754	0.10	0.05
259	151	0.25	0.15
130	75.4	0.40	0.22
51.9	30.2	0.62	0.41
25.9	15.1	0.90	0.73
13.0	7.54	1.40	1.00
5.19	3.02	2.40	1.75

concentrations the distance to a given concentration is cut in half as roughness increases from 0.03 to 0.3 m. At small concentrations, the distance is cut by a lesser amount (about 30%) with the same change in roughness. These results will be compared with the SLAB and DEGADIS model predictions in later sections.

SLAB MODEL

The SLAB model is valid for gaseous emissions from area sources, and can account for departures of plume temperature and density from ambient. The SLAB model is applied only to Case 1: 21% of chlorine emitted as flashed vapor and 79% of chlorine spilled as a liquid, which is subsequently emitted by evaporation from a pool. Each of these two sources is modeled separately and the results summed, thus ignoring the additive effects of the density perturbations. We point out that some other investigators combine the sources prior to applying the dispersion model, thus requiring only a single model run. Also, some investigators have devised ways to apply the SLAB model to Case 2, which involves aerosols carried by the gas plume, but we have not attempted this calculation because the original model code has no explicit guidance regarding aerosols.

The model automatically accounts for variable input values of roughness lengths, wind speeds, and averaging times. Stability class is parameterized by an inverse Monin–Obukhov length, $1/L$, calculated by a preprocessor:

	$z_0 = 0.03\,m$	$z_0 = 0.30\,m$
Class F, $u = 2$ m/s	$1/L = 0.07025$ m^{-1}	$1/L = 0.04733$ m^{-1}
Class D, $u = 2$ or 5 m/s	$1/L = 0$	$1/L = 0$

In our SLAB runs, concentrations are calculated to a maximum downwind distance of 3 km (see example 6 in the Appendix).

The predicted concentrations listed in Table 3-8 for the SLAB model will be compared with both the CAMEO model and DEGADIS model predicted concentrations later. In general the SLAB model predictions are less than the CAMEO model predictions by a factor of two to five for the stable case and are approximately equal to the CAMEO predictions for the neutral case. The table shows that the contribution from the flashed vapor component is a factor of about three to four less than the contribution from the pool source, in agreement with the difference in the magnitude of their source emissions ($Q_{pool}/Q_{vapor} = 3.8$). In all cases, an increase in roughness from 0.03 to 0.3 m causes a decrease in concentration of about 30 to 40%. Changes in stability class and wind speed are seen to have less of an influence close to the source than farther away, due to the size of the pool area source.

DEGADIS MODEL

The DEGADIS model was run for all combinations of stability class, wind speed, roughness, and averaging times, for Case 1 (no aerosol entrainment) and Case 2

TABLE 3-8
SLAB Model Concentration Predictions (ppm) for Scenario 4, Case 1 (No Aerosol Entrainment)[a]

Downwind distance (km)	1 min averaging time						15 min averaging time					
	F, 2 m/s $z_0 =$		D, 2 m/s $z_0 =$		D, 5 m/s $z_0 =$		F, 2 m/s $z_0 =$		D, 2 m/s $z_0 =$		D, 5 m/s $z_0 =$	
	0.03 m	0.3 m	0.03 m	0.3 m	0.03 m	0.3 m	0.03 m	0.3 m	0.03 m	0.3 m	0.03 m	0.3 m
Flashed vapor												
0.05	3050.0	1880.0	2440.0	1370.0	1010.0	633.0	2900.0	1750.0	1910.0	1060.0	572.0	359.0
0.09	1450.0	865.0	988.0	540.0	366.0	227.0	1280.0	742.0	685.0	375.0	204.0	128.0
0.19	492.0	280.0	254.0	144.0	90.5	57.4	384.0	212.0	157.0	90.5	50.6	32.3
0.40	149.0	82.8	62.3	38.5	22.9	15.1	104.0	56.5	36.7	23.0	13.0	8.56
0.57	79.6	44.9	31.6	20.4	11.8	7.92	53.0	29.5	18.4	12.0	6.71	4.52
0.83	42.7	24.8	16.4	11.0	6.20	4.25	27.5	15.8	9.48	6.37	3.55	2.44
0.99	31.5	18.6	11.9	8.09	4.53	3.14	20.0	11.7	6.89	4.69	2.60	1.80
1.44	17.5	10.7	6.44	4.49	2.48	1.74	10.9	6.61	3.70	2.59	1.42	1.00
2.08	10.1	6.30	3.57	2.54	1.38	0.99	6.18	3.85	2.05	1.46	0.79	0.57
3.00	5.97	3.82	2.03	1.47	0.79	0.57	3.63	2.31	1.16	0.85	0.46	0.33
Pool source												
0.05	8400.0	4940.0	7600.0	4000.0	3100.0	1850.0	8470.0	4950.0	7210.0	3720.0	2060.0	1250.0
0.10	3390.0	1940.0	2810.0	1400.0	1040.0	617.0	3380.0	1910.0	2450.0	1194.0	615.0	372.0
0.19	1270.0	710.0	780.0	441.0	331.0	203.0	1200.0	655.0	669.0	336.0	189.0	117.0
0.38	446.0	246.0	244.0	138.0	104.0	66.7	384.0	206.0	166.0	94.7	58.7	38.0
0.57	230.0	130.0	114.0	69.4	45.0	29.8	187.0	103.0	74.3	45.6	25.4	17.0
0.75	148.0	85.6	70.2	44.4	28.1	19.0	115.0	65.3	44.5	28.4	15.9	10.8
0.99	96.0	56.9	43.6	28.6	17.3	11.9	72.2	42.1	27.1	17.9	9.84	6.80
1.50	51.3	31.6	22.0	15.0	8.20	5.80	37.0	22.4	13.4	9.21	4.69	3.32
1.98	34.4	21.6	14.2	9.90	6.12	4.40	24.3	15.1	8.53	6.00	3.50	2.50
3.00	19.4	12.5	7.52	5.40	3.00	2.18	13.4	8.52	4.48	3.23	1.72	1.25
Total flashed vapor and pool source												
0.57	310.0	175.0	146.0	89.8	56.2	37.7	240.0	133.0	92.7	57.6	32.1	21.5
0.99	128.0	75.5	55.5	36.7	21.8	15.0	92.2	53.8	34.0	22.6	12.4	8.60
3.00	25.4	16.3	9.55	6.87	3.79	2.75	17.0	10.8	5.64	4.08	2.18	1.58

[a]Maximum ground level centerline concentrations are listed for the flashed vapor and the pool source components, and then summed at three distances in the last part.

(50% aerosol entrainment), resulting in 48 separate printouts. To save space, concentrations at only one specific distance are given in the following tables for each case, for comparison with CAMEO and SLAB model runs. Two examples of variation of concentration with downwind distance for the three models are given.

The "δ" factors (leading coefficients in the σ_y power law formulas) were changed in the DEGADIS version 1.4 model to account for variable averaging times:

	1 min averaging time	15 min averaging time
Class F	$\delta = 0.0404$	$\delta = 0.0694$
Class D	$\delta = 0.0820$	$\delta = 0.141$

For Case 1, the DEGADIS model was run in "nonisothermal" mode, which accounts for effects of heat transfer from the underlying surface. The vapor source was arbitrarily assumed to be emitted from a surface area of $1\,m^2$, corresponding to a radius of 0.56 m. The following input parameters were used:

Vapor: $Q = 0.2457\,kg/s$ radius = 0.56 m
Pool: $Q = 0.9243\,kg/s$ radius = 4.8 m ($u = 2$ m/s)
                                            radius = 3.1 m ($u = 5$ m/s)

For Case 2, the DEGADIS model was run in "isothermal" mode, with the initial vapor density changed to $8.188\,kg/m^3$ to reflect the aerosol content. The following input parameters were used:

Vapor: $Q = 0.708\,kg/s$ radius = 0.56 m
Pool: $Q = 0.462\,kg/s$ radius = 3.48 m ($u = 2$ m/s)
                                             radius = 2.19 m ($u = 5$ m/s)

See Sections 4 and 5 of the Appendix for the output for an example from each case.

Table 3-9 lists the predicted concentrations at a downwind distance of 1.0 km for the three models and the various conditions that were run. In the case of the SLAB and DEGADIS models, the contributions of the pool and vapor components to the total concentration are given. DEGADIS was the only model that was run for the 50% aerosol input assumption (Case 2). Although the relative contributions of the pool and vapor components in Cases 1 and 2 differ, the results in the table show that the total concentration predicted by DEGADIS differs by less than ±5% between the two cases. If transient behavior of the source had been modeled, the results of the comparison would have been different.

The SLAB and DEGADIS model predictions are always within about ±20% of each other in the table. Unlike Scenario 2 (ammonia), where the DEGADIS model predicted a large source blanket (i.e., layer with 100% vapor concentration) and hence high concentrations in the near field, in this scenario

TABLE 3-9

Comparison of CAMEO, SLAB, and DEGADIS Model Concentration Predictions at a Downwind Distance of 1.0 km for Scenario 4 (Chlorine)[a]

	1 min averaging time						15 min averaging time					
	CAMEO $z_0=$		SLAB $z_0=$		DEGADIS $z_0=$		CAMEO $z_0=$		SLAB $z_0=$		DEGADIS $z_0=$	
	0.03	0.3 (ppm)	0.03	0.3 (ppm)	0.03	0.3 (ppm)	0.03	0.3 (ppm)	0.03	0.3 (ppm)	0.03	0.3 (ppm)
Case 1 (no aerosol)												
F, 2 m/s												
P			96.0	56.9	118.0	57.1			72.2	42.1	96.8	42.8
V			31.5	18.6	39.3	19.3			20.0	11.7	25.3	12.6
T	358.0	179.0	128.0	75.5	157.0	76.4	214.0	102.0	92.2	53.8	122.0	55.4
D, 2 m/s												
P			43.6	28.6	61.4	29.8			27.1	17.9	36.5	17.2
V			11.9	8.09	14.7	7.71			6.89	4.69	8.77	4.56
T	56.1	38.6	55.5	36.7	76.1	37.5	32.3	20.0	34.0	22.6	45.2	21.7
D, 5 m/s												
P			17.3	11.9	21.4	11.6			9.84	6.80	12.3	6.66
V			4.53	3.14	5.61	3.16			2.60	1.80	3.16	1.75
T	21.4	13.0	21.8	15.0	27.0	14.7	13.0	7.71	12.4	8.60	15.4	8.42
Case 2 (50% aerosol)	(g/m^3)						(g/m^3)					
F, 2 m/s												
P			0.204		0.098				0.139		0.064	
V			0.224		0.104				0.200		0.088	
T			0.428		0.202				0.339		0.152	
D, 2 m/s												
P			0.085		0.041				0.049		0.024	
V			0.135		0.061				0.080		0.038	
T			0.220		0.102				0.129		0.062	
D, 5 m/s												
P			0.030		0.017				0.017		0.0095	
V			0.046		0.025				0.027		0.0145	
T			0.076		0.042				0.044		0.0240	

[a]Case 1 refers to no aerosol entrainment and Case 2 refers to 50% of liquid aerosol entrainment. Pool (P) and vapor (V) contributions are shown for SLAB and DEGADIS. The total (T) equals the sum of P plus V.

the source blanket is predicted to be relatively small, with a half-width of less than 10 m, and little influence on concentrations at distance beyond about 50 m.

The CAMEO model predictions in Table 3-9 are close to the SLAB and DEGADIS model predictions for stability class D (nearly neutral) conditions. During class F conditions the CAMEO model predictions are two to four times the predictions of the other models. This can be seen in Figures 3-9 and 3-10, where the predictions of the three models are plotted as a function of downwind distance for two extreme situations.

Figure 3-9:   Class D, $u = 5$ m/s, $z_0 = 0.3$ m,   15 min averaging time
Figure 3-10:   Class F, $u = 2$ m/s, $z_0 = 0.3$ m,   1 min averaging time

Figure 3-9 shows the lowest concentrations of all the conditions studied, and Figure 3-10 shows one of the highest. The SLAB and DEGADIS model curves are seen to be within a factor of two of each other at all points in the two figures, and at distances beyond 0.5 km the two curves are within ±10% of each other. The CAMEO model curve is slightly below the others for the class D figure, and a factor of two to four above the others for the class F figure. The slopes of all the curves are quite similar, even though their magnitudes may be slightly different.

Since dense-gas effects are more important near the source, the biggest differences between the CAMEO model and the other models are expected at

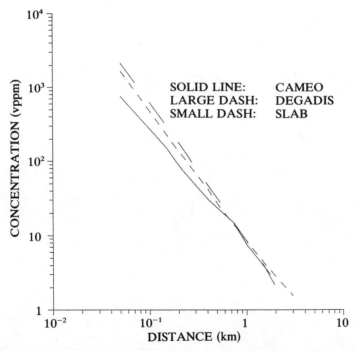

Figure 3-9. *Comparison of CAMEO, SLAB, and DEGADIS model predictions for Scenario 4 (chlorine), for Case 1 (no entrained aerosol) and for class D, $u = 5$ m/s, $z_0 = 0.3$ m, 15 min averaging time. For the SLAB and DEGADIS models, the vapor and pool contributions have been summed.*

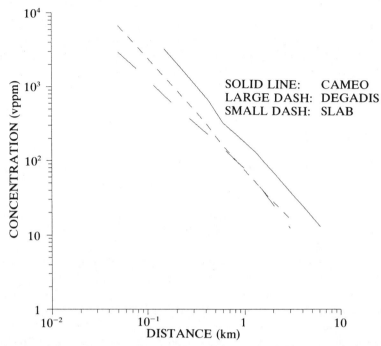

Figure 3-10. *Comparison of CAMEO, SLAB, and DEGADIS model predictions for Scenario 4 (chlorine) for Case 1 (no entrained aerosol) and for Class F, u = 2 m/s, z_0 = 0.3 m, 1 min averaging time. For the SLAB and DEGADIS models the vapor and pool contributions have been summed.*

distances of 50 m or less. But we hesitate to make comparisons in that range because of the unrealistic source geometry assumptions that must be made to apply the SLAB and DEGADIS models to a jet source. Revised versions of these models will permit the initial jet to be explicitly modeled.

NOTES

NOTES

3.5 Scenario 5: Liquid Spill of Acetone

A spill of liquid acetone from a 2-in. nozzle connection near the base of a cylindrical tank of height 25 ft and diameter 25 ft occurs over a 30-min period into a concrete diked area of height 3 ft and diameter 72 ft. Figure 3-11 contains a schematic drawing of this case. The chemical is stored at ambient pressure and temperature (303°K). Maximum ground level concentrations for 1 and 30 min averaging times are desired as a function of downwind distance. Evaporation rate is calculated using the SPILLS model. Downwind transport and dispersion is calculated using the SPILLS and DEGADIS models.

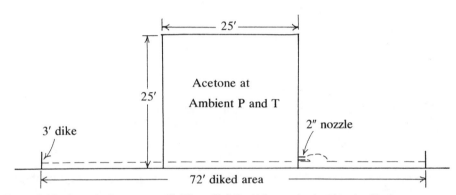

Figure 3-11. *Scenario 5: acetone spill. The spilled liquid is contained within the diked area.*

3.5.1 *Source Emissions*

Acetone is a slowly evaporating chemical that will spill as a liquid from the tank, with no flashing.

Properties of Acetone:
Molecular weight	$M_g = 58.1$ kg/kg-mol
Density of liquid	$\rho_l = 791$ kg/m^3
Density of gas	$\rho_g = 2.34$ kg/m^3 at 303°K

Tank Data:
Tank diameter	$= 7.62$ m
Tank height	$H = 7.62$ m
Nozzle diameter	$D = 0.051$ m
stored at ambient P and T	

Ambient Meteorology:
Pressure	$p_a = 1.013 \times 10^5$ N/m^2
Temperature	$T_a = 303$°K

Relative humidity	= 50%
Class D	$u = 2$ m/s
Class D	$u = 5$ m/s
Class F	$u = 2$ m/s

Duration of Spill: 30 min

The Bernoulli equation (2-3) can be used for calculating the mass flux, Q_l, of liquid flow from a nozzle:

$$Q_l = C_D A(\rho_l)[2(\Delta p)/(\rho_l) + 2gH]^{0.5}$$

where $C_D = 0.6$
$A = \pi D^2/4 = 2.04 \times 10^{-3}$ m^2
$\rho_l = 780$ kg/m^3
$\Delta p = 0.0$
$H = 7$ m

Therefore, $Q_l = 11.2$ kg/s.

The time dependence of the leak rate is described by Eq. (2-2):

$$Q_l(t) = 0.6 A\rho_l(2gH_0)^{0.5} - t(0.6A)^2\rho_l g/a$$

After integration over time t, the total mass released, M, can be calculated:

$$M = \rho_l C_D At[(2gH_0)^{0.5} - (C_D Ag)/(2a)t]$$

Using $t = 30$ min $= 1800$ s, and letting A equal the nozzle area (2.043×10^{-3} m^2) and a equal the tank area, then the mass released, M, is calculated to be 19,745 kg.

The SPILLS model can be used to calculate evaporation from the acetone pool. This model uses Eqs. (2-10) through (2-12) for slow evaporation, which can be solved as a hand-check on the model output:

$$Q_e = k_g A p_s M/R^* T$$

where $k_g = D_m N_{Sh}/d$ and $N_{Sh} = 0.037(k_m/D_m)^{1/3}[(ud/k_m)^{0.8} - 15200]$ [Eq. (2.10)]
$u = 2$ m/s
$d = 21.95$ m (diameter of diked area)
$A = $ area of liquid pool $= 332.7$ m^2 (subtracting tank area)
$p_s = $ (vapor pressure) $= 3.73 \times 10^4$ N/m^2 at $T = 303°$K
$M = 58$ kg/kg-mol
$R^* = 8314$ J/kg-mol/°K
$T = 303°$K
$k_m/D_m = 1$
$k_m = 0.11 \times 10^{-4}$ m^2 s^{-1}

Therefore, $Q_e = 0.96$ kg/s for the 2 m/s example and 2.10 kg/s for the 5 m/s example. With these evaporation rates, several hours are required to completely evaporate the spilled liquid. The SPILLS model gives $Q_e = 0.84$ kg/s for the 2 m/s example and 1.88 kg/s for the 5 m/s example. The hand-check is within 15% of the SPILLS model output, which is reasonable considering that the hand-check employed only rough estimates of K_m and D_m.

3.5.2 *Transport and Dispersion Calculated Using the SPILLS and DEGADIS Models*

Both the SPILLS and DEGADIS models are appropriate for calculating transport and dispersion from an evaporating pool. The major difference between the models is that SPILLS cannot handle dense gases whereas DEGADIS can. To determine whether dense-gas effects are important in this scenario, the initial relative excess density $(\rho_g - \rho_a)/\rho_a$ at the source should be calculated. The source emission rate is about 2 kg/s and the acetone gas can be assumed to be uniformly spread over a cross section of height, $h = 1$ m, and width, $W = 22$ m, at the downwind edge of the spill pool. Since the wind speed is 5 m/s, the concentration of acetone gas at that location is $C = Q_e/uhW = 0.018$ kg/m^3. Consequently the initial relative excess density, $(\rho_g - \rho_a)/\rho_a$, is on the order of 0.01, which is on the borderline of the criterion in the DEGADIS model for considering dense-gas effects. We conclude that the DEGADIS model is really not called for in this scenario, but make calculations with the model anyway to compare its predictions with the SPILLS model predictions. Presumably the DEGADIS model should agree very closely with a Gaussian model in the limit of zero excess density.

SPILLS

This model does not account for varying roughness lengths or averaging times. Because it is based on the EPA σ_y and σ_z curves, it is most appropriate for roughness lengths of about 0.1 m and averaging times of about 10 min. To convert the SPILLS model 10 min averaged concentrations to 1 and 30 min averaged concentrations, the following formulas are used [from Eq. (2-21)]:

$$C(t_a = 1 \text{ min}) = 1.585 C(\text{SPILLS}; \ t_a = 10 \text{ min})$$
$$C(t_a = 30 \text{ min}) = 0.803 C(\text{SPILLS}; \ t_a = 10 \text{ min})$$

TABLE 3-10
SPILLS Model Predictions of Concentrations (vppm) as a Function of Downwind Distance, x, for Scenario 5

	$t_a = 1 \ min$			$t_a = 30 \ min$		
	Class D	*Class F*	*Class D*	*Class D*	*Class F*	*Class D*
x (km)	*u = 2 m/s*	*u = 2 m/s*	*u = 5 m/s*	*u = 2 m/s*	*u = 2 m/s*	*u = 5 m/s*
0.1	1650.0	4987.0	1475.0	836.0	2530.0	748.0
0.2	562.0	1959.0	502.0	285.0	993.0	255.0
0.4	182.0	714.0	163.0	92.0	362.0	82.6
0.8	57.8	246.0	51.8	29.3	125.0	26.2
1.6	19.5	87.8	17.4	9.89	44.5	8.82
3.2	6.8	32.6	6.0	3.43	16.5	3.06

The SPILLS model was run for the three combinations of wind speed and stability class [$u = 2$ m/s, class D; $u = 2$ m/s, class F (output for this case is listed in Section 9 of the Appendix); $u = 5$ m/s, class D] that were specified in this scenario. Results for six downwind distances are given in Table 3-10.

It is seen that the maximum concentrations in Table 3-10 are occurring for the stable, light-wind case. The predicted concentrations for the two wind speeds in class D are very similar (within ±10%), since the increase in evaporative emission due to the 5 m/s wind speed is nearly counterbalanced by the increased dilution. In all cases the concentration is dropping by a factor of about three as downwind distance doubles.

DEGADIS

Because the DEGADIS model can account for the differences between the two roughness lengths, 0.03 and 0.3 m, and between the two averaging times, 1 and 30 min, there are 12 separate model runs that must be made. An example of the output from one of these runs is given in Section 8 of the Appendix. Model predictions at the same six distances as the SPILLS model predictions in Table 3-10 are given in Table 3-11. These runs are generated using the following input parameters specific to DEGADIS:

Heat capacity constant = 0   (no ΔT effects)
Heat capacity power = 1
Lowest concentration of interest = 2.3×10^{-5} kg/m^3 or $mf_{LFL} = 2.0 \times 10^{-5}$
$Q = 0.8419$ kg/s ($u = 2$ m/s)
$Q = 1.882$ kg/s ($u = 5$ m/s)
Source radius = 10.29 m (area = 332.7 m^2)

For averaging times of 1 and 30 min, change the "δ" coefficient in the σ_y formula to the values in the following table:

	$t_a = 1$ min	$t_a = 30$ min
Stability class = F	$\delta = 0.0404$	$\delta = 0.0797$
Stability class = D	$\delta = 0.0820$	$\delta = 0.162$

Comparing the SPILLS model predictions in Table 3-10 with the DEGADIS model predictions in Table 3-11, the following conclusions can be reached:

• In most conditions, the SPILLS model predictions (valid for a roughness of about 0.1 m) fall in between the DEGADIS model predictions for roughnesses of 0.03 and 0.3 m. The biggest difference occurs for the class F runs at distances of 0.4 km or less, where the dense-gas effects are accounted for by the DEGADIS model. However, these data suggest that dense-gas effects are not important in this scenario for neutral conditions.
• The averaging time correction in the DEGADIS model becomes less important near the source, because it does not apply to the "source blanket" portion of the lateral dispersion. This correction is applied in the DEGADIS model only to the turbulent part of lateral dispersion, σ_y. In

TABLE 3-11

DEGADIS Model Predictions of Concentrations (vppm) as a Function of Downwind Distance, x, for Scenario 5

x (km)	t_a = 1 min						t_a = 30 min					
	Class D u = 2 m/s		Class F u = 2 m/s		Class D u = 5 m/s		Class D u = 2 m/s		Class F u = 2 m/s		Class D u = 5 m/s	
	z_0 = 0.03 m	z_0 = 0.3 m	z_0 = 0.03 m	z_0 = 0.3 m	z_0 = 0.03 m	z_0 = 0.3 m	z_0 = 0.03 m	z_0 = 0.3 m	z_0 = 0.03 m	z_0 = 0.3 m	z_0 = 0.03 m	z_0 = 0.3 m
0.1	3570.0	1391.0	4156.0	3531.0	1930.0	993.0	2945.0	1156.0	4156.0	3647.0	1609.0	830.0
0.2	1434.0	561.0	1708.0	1725.0	796.0	412.0	830.0	342.0	1708.0	1378.0	492.0	258.0
0.4	419.0	175.0	749.0	655.0	257.0	134.0	216.0	95.9	698.0	372.0	146.0	76.6
0.8	111.0	49.7	309.0	178.0	74.9	40.0	53.9	25.7	185.0	93.3	40.0	21.5
1.6	27.6	13.5	87.7	45.4	21.1	11.2	25.7		44.9	14.5	11.0	
3.2	22.2		22.2	11.8					11.2			

65

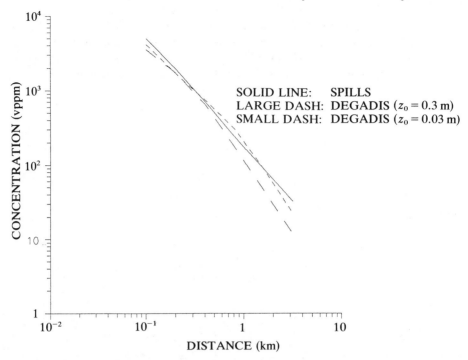

Figure 3-12. *Comparison of SPILLS and DEGADIS model predictions for stability class F and a 2 m/s wind speed, for a 1-min averaging time (Scenario 5). Data are from Tables 3-10 and 3-11.*

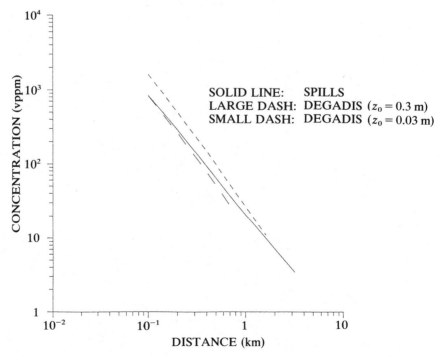

Figure 3-13. *Comparison of SPILLS and DEGADIS model predictions for stability class D and a 5 m/s wind speed, for a 30-min averaging time (Scenario 5). Data are from Tables 3-10 and 3-11.*

contrast, this correction is equally important at all distances for the SPILLS model runs, since it is applied to the total cloud width.

- At large distances downwind the DEGADIS model predictions suggest that the concentration is reduced by a factor of about two as the roughness increases from 0.03 to 0.3 m.

The model predictions are further compared in Figures 3-12 and 3-13 for two of the six conditions in the tables. Visual inspection of the curves drawn on these figures verifies that the two model predictions are indeed quite similar over this range of conditions. This conclusion supports the calculation made earlier that the initial excess relative density of the acetone gas is only about 0.01, which is on the borderline for use of the DEGADIS model.

NOTES

NOTES

APPENDIX

Examples of Output
from Computer Models

1. Scenario 1: DEGADIS

Stability Class = F
Wind Speed $u = 2$ m/s
Roughness Length $z_0 = 0.3$ m
Averaging time $t_a = 10$ s

0   TITLE BLOCK

SC=F   WS=2   ZO=0.3   PURE nC4

              Wind velocity at reference height   2.00   m/s
              Reference height   10.00   m
0   Surface roughness length   0.300   m
0   Pasquill Stability class   F
0   Monin-Obukhov length   21.1   m
              Gaussian distribution constants   Delta   0.02820   m
                                                   Beta   0.90000
0   Wind velocity power law constant   Alpha   0.53216
              Friction velocity   0.12155   m/s
0   Ambient Temperature   303.15   K
              Ambient Pressure   1.000   atm
              Ambient Absolute Humidity   1.389E-02   kg/kg BDA
              Ambient Relative Humidity   50.00   %

       Input: Mole fraction   CONCENTRATION OF C   GAS DENSITY
                                                 kg/m**3   kg/m**3
                             0.00000   0.00000   1.15593
                             1.00000   2.33800   2.33800

0   Specified Gas Properties:

              Molecular weight: 58.120
              Storage temperature: 303.00   K
              Density at storage temperature and ambient pressure: 2.3380   kg/m**3
              Mean heat capacity constant: ********   [NOT USED]
              Mean heat capacity power: ********   [NOT USED]
              Upper mole fraction contour: 1.90000E-02
              Lower mole fraction contour: 4.75000E-03
              Height for isopleths: 0.00000E-01m

     Source input data points

              Initial mass in cloud: 0.00000E-01

	TIME	SOURCE STRENGTH	SOURCE RADIUS
	s	kg/s	m
	0.00000E-01	15.000	6.7400
	6023.0	15.000	6.7400
	6024.0	0.00000E-01	0.00000E-01
	6025.0	0.00000E-01	0.00000E-01

0   Calculation procedure for ALPHA: 1
0   Entrainment prescription for PHI: 3
0   Layer thickness ratio used for average depth: 2.1500
0   Air entrainment coefficient used: 0.590
0   Gravity slumping velocity coefficient used: 1.150
0   Isothermal calculation
0   Heat transfer not included
0   Water transfer not included

*****   CALCULATED SOURCE PARAMETERS   *****

Time	Gas Radius	Height	Qstar	SZ(x=L/2.)	Mole frac C	Density	Rich No.
sec	m	m	kg/m**2/s	m		kg/m**3	
0.000000E-01	6.74000	1.100000E-05	6.629437E-03	0.205248	1.00000	2.33800	0.756144
0.160000	6.77187	6.714722E-03	6.597587E-03	0.205812	0.995705	2.33292	0.756144
0.320000	6.82969	1.329165E-02	6.540590E-03	0.206829	0.988033	2.32385	0.756144
0.480000	6.90382	1.969247E-02	6.468959E-03	0.208126	0.978405	2.31247	0.756144
0.640000	6.99056	2.588721E-02	6.387129E-03	0.209631	0.967425	2.29949	0.756144
0.960000	7.19344	3.758999E-02	6.203695E-03	0.213101	0.942883	2.27048	0.756144
1.28000	7.42642	4.832807E-02	6.005805E-03	0.217004	0.916520	2.23932	0.756144
2.56000	8.53786	8.211842E-02	5.214626E-03	0.234515	0.812293	2.11612	0.756144
3.84000	9.78837	0.104335	4.544797E-03	0.252317	0.725548	2.01358	0.756144
5.12000	11.0848	0.118933	4.013778E-03	0.269011	0.657780	1.93347	0.756144
7.68000	13.6820	0.135063	3.258825E-03	0.298160	0.563041	1.82148	0.756144
10.2400	16.2134	0.142068	2.759782E-03	0.322277	0.501548	1.74879	0.756144
15.3600	21.0191	0.144710	2.147521E-03	0.359578	0.427782	1.66160	0.756144
20.4800	25.5118	0.141412	1.785868E-03	0.387126	0.385596	1.61173	0.756144
25.6000	29.7453	0.136025	1.545859E-03	0.408267	0.358710	1.57995	0.756144
30.7200	33.7632	0.129891	1.374114E-03	0.424888	0.340459	1.55838	0.756144
35.8400	37.5981	0.123557	1.244640E-03	0.438139	0.327619	1.54320	0.756144
40.9600	41.2745	0.117269	1.143251E-03	0.448773	0.318438	1.53235	0.756144
51.2000	48.2230	0.105225	9.941598E-04	0.464113	0.307286	1.51916	0.756144
61.4400	54.7167	9.410075E-02	8.894159E-04	0.473579	0.302427	1.51342	0.756144
76.8000	63.7535	7.917132E-02	7.797864E-04	0.479958	0.302672	1.51371	0.756144
97.2800	74.7123	6.220831E-02	6.835677E-04	0.478126	0.312762	1.52564	0.756144
112.640	81.5729	5.223082E-02	6.377292E-04	0.470792	0.326220	1.54155	0.756144
168.960	84.2073	4.729476E-02	6.321905E-04	0.441836	0.367939	1.59086	0.756144
245.760	86.1715	4.392009E-02	6.272180E-04	0.423530	0.398588	1.62709	0.756144
286.720	86.7030	4.290839E-02	6.260978E-04	0.422121	0.408195	1.63845	0.756144

0Source strength [kg/s] :   15.000   Equivalent Primary source radius [m] :   6.7400
 Equivalent Primary source length [m] :   11.946   Equivalent Primary source width [m] :   11.946

75

Secondary source concentration [kg/m**3] :   0.95457   Secondary source SZ [m] :   0.42212

Contaminant flux rate: 6.35145E-04

Secondary source mass fractions... contaminant: 0.582604   air: 0.41168
          Enthalpy: 0.00000E-01   Density: 1.6384
Secondary source length [m] :   153.68   Secondary source half-width [m] :   76.839

0 Distance	Mole Fraction	Concentration	Density	Temperature	Half Width	Sz	Sy	Width at z= 0.00 m to: 0.475 mole%	1.90 mole%
(m)		(kg/m**3)	(kg/m**3)	(K)	(m)	(m)	(m)	(m)	(m)
76.8	0.408	0.955	1.64	303.	76.8	0.422	3.000E-02	76.9	76.9
77.6	0.391	0.915	1.62	303.	79.8	0.418	1.47	82.9	82.3
78.4	0.375	0.877	1.60	303.	83.4	0.416	2.11	87.8	87.0
80.0	0.345	0.807	1.56	303.	90.5	0.416	3.05	96.8	95.7
81.6	0.314	0.733	1.53	303.	97.5	0.420	3.81	105.	104.
83.2	0.284	0.664	1.49	303.	104.	0.429	4.48	113.	112.
86.4	0.229	0.535	1.43	303.	117.	0.456	5.68	128.	126.
89.6	0.184	0.431	1.37	303.	128.	0.494	6.75	141.	138.
96.0	0.120	0.282	1.30	303.	147.	0.594	8.68	162.	158.
102.	8.178E-02	0.191	1.25	303.	162.	0.716	10.4	179.	174.
115.	4.277E-02	0.100	1.21	303.	184.	0.998	13.5	204.	196.
128.	2.573E-02	6.018E-02	1.19	303.	201.	1.31	16.2	222.	210.
141.	1.738E-02	4.067E-02	1.18	303.	201.	1.68	18.7	223.	
154.	1.280E-02	2.994E-02	1.17	303.	199.	2.05	20.9	220.	
166.	9.985E-03	2.336E-02	1.17	303.	198.	2.41	22.8	217.	
179.	8.105E-03	1.896E-02	1.17	303.	196.	2.76	24.6	214.	
192.	6.772E-03	1.584E-02	1.16	303.	194.	3.11	26.3	210.	
205.	5.785E-03	1.353E-02	1.16	303.	193.	3.44	27.9	205.	
218.	5.028E-03	1.176E-02	1.16	303.	192.	3.77	29.4	199.	
230.	4.432E-03	1.037E-02	1.16	303.	190.	4.09	30.8		
243.	3.951E-03	9.242E-03	1.16	303.	189.	4.41	32.2		
256.	3.556E-03	8.319E-03	1.16	303.	188.	4.73	33.5		
269.	3.227E-03	7.549E-03	1.16	303.	187.	5.04	34.7		
282.	2.949E-03	6.899E-03	1.16	303.	186.	5.34	35.9		
294.	2.712E-03	6.344E-03	1.16	303.	185.	5.64	37.1		
307.	2.507E-03	5.864E-03	1.16	303.	184.	5.94	38.2		

For the UFL of   1.9000   mole percent, and the LFL of   0.47500   mole percent:

The mass of contaminant between the UFL and LFL is: -217.79   kg.
The mass of contaminant above the LFL is: 0.00000E-01 kg.

2. Scenario 1: SLAB

Stability Class = F
Wind Speed $u = 2$ m/s
Roughness Length $z_0 = 0.3$ m
Averaging time $t_a = 10$ s

```
problem input

    wms   =       .058
    cps   =   1715.000
    ts    =    303.000
    qs    =     15.000
    as    =    143.000
    avt   =     10.000
    xffm  =   1000.000
    zp(2) =      1.000
    zp(3) =       .000
    zp(4) =       .000
```

release gas properties

molecular weight of source gas (kg)	- wms =	.5812E-01
heat capacity at const. p. (j/kg-k)	- cps =	.1715E+04
temperature of source gas (k)	- ts =	.3030E+03
density of source gas (kg/m3)	- rhos =	.2338E+01

spill characteristics

mass source rate (kg/s)	- qs =	.1500E+02
source area (m2)	- as =	.1430E+03
vapor injection velocity (evaporation rate) (m/s)	- ws =	.4487E-01
source half width -.5*((qqs/ws)**.5) (m)	- bs =	.5979E+01

field parameters

concentration averaging time (s)	- avt =	.1000E+02
maximum downwind distrace (m)	- xffm =	.1000E+04
concentration measurement height (m)	- zp(1)=	.0000E+00
	- zp(2)=	.1000E+01
	- zp(3)=	.0000E+00
	- zp(4)=	.0000E+00

ambient meteorological properties

molecular weight of air (kg)	- wma =	.2896E-01
heat capacity of air at const. p. (j/kg-k)	- cpa =	.1006E+04
density of ambient air (kg/m3)	- rhoa =	.1165E+01
ambient measurement height (m)	- za =	.1000E+02
ambient atmospheric pressure (atm)	- pa =	.1000E+01
ambient wind speed (m/s)	- ua =	.2000E+01
ambient temperature (k)	- ta =	.3030E+03
ambient friction velocity (m/s)	- uastr =	.2191E+00
inverse monin-obukhov length (1/m)	- ala =	.4733E-01
surface roughness height (m)	- z0 =	.3000E+00

additional parameters

acceleration of gravity (m/s2)	- grav =	.9807E+01

```
gas constant (m3 -atm/mol- k)                          - rr   =   .8206E-04
von karman constant                                    - xk   =   .4100E+00




source region

         [INTERMEDIATE OUTPUT REMOVED]

volume concentration of (x,z)

      x        z=  .00     z= 1.00     z=  .00     z=  .00
  -.1444E+02   .0000E+00   .0000E+00   .0000E+00   .0000E+00
  -.1155E+02   .2049E-01   .2799E-02   .0000E+00   .0000E+00
  -.8662E+01   .3322E-01   .9767E-02   .0000E+00   .0000E+00
  -.5775E+01   .4426E-01   .1788E-01   .0000E+00   .0000E+00
  -.2887E+01   .5415E-01   .2638E-01   .0000E+00   .0000E+00
  -.2623E-05   .6314E-01   .3490E-01   .0000E+00   .0000E+00
   .2887E+01   .7149E-01   .4318E-01   .0000E+00   .0000E+00
   .5775E+01   .7945E-01   .5112E-01   .0000E+00   .0000E+00
   .8662E+01   .8719E-01   .5870E-01   .0000E+00   .0000E+00
   .1155E+02   .9471E-01   .6602E-01   .0000E+00   .0000E+00
   .1444E+02   .1018E+00   .7513E-01   .0000E+00   .0000E+00




      far field



         input source parameters      effective source parameters
             bs =  .5979E+01              bse =  .1444E+02
             ws =  .4487E-01              wse =  .7697E-02

         [INTERMEDIATE OUTPUT REMOVED]

volume concentration of (x,z)

      x        z=  .00     z= 1.00     z=  .00     z=  .00
  -.1444E+02   .0000E+00   .0000E+00   .0000E+00   .0000E+00
  -.1155E+02   .2049E-01   .2799E-02   .0000E+00   .0000E+00
  -.8662E+01   .3322E-01   .9767E-02   .0000E+00   .0000E+00
  -.5775E+01   .4426E-01   .1788E-01   .0000E+00   .0000E+00
  -.2887E+01   .5415E-01   .2638E-01   .0000E+00   .0000E+00
  -.2623E-05   .6314E-01   .3490E-01   .0000E+00   .0000E+00
   .2887E+01   .7149E-01   .4318E-01   .0000E+00   .0000E+00
   .5775E+01   .7945E-01   .5112E-01   .0000E+00   .0000E+00
   .8662E+01   .8719E-01   .5870E-01   .0000E+00   .0000E+00
   .1155E+02   .9471E-01   .6602E-01   .0000E+00   .0000E+00
   .1444E+02   .1018E+00   .7513E-01   .0000E+00   .0000E+00
   .1540E+02   .1009E+00   .6889E-01   .0000E+00   .0000E+00
   .1645E+02   .9965E-01   .6505E-01   .0000E+00   .0000E+00
   .1761E+02   .9814E-01   .6143E-01   .0000E+00   .0000E+00
   .1887E+02   .9632E-01   .5788E-01   .0000E+00   .0000E+00
   .2026E+02   .9414E-01   .5436E-01   .0000E+00   .0000E+00
   .2178E+02   .9158E-01   .5091E-01   .0000E+00   .0000E+00
   .2345E+02   .8863E-01   .4756E-01   .0000E+00   .0000E+00
   .2528E+02   .8530E-01   .4438E-01   .0000E+00   .0000E+00
   .2729E+02   .8161E-01   .4140E-01   .0000E+00   .0000E+00
   .2949E+02   .7758E-01   .3866E-01   .0000E+00   .0000E+00
   .3190E+02   .7328E-01   .3618E-01   .0000E+00   .0000E+00
```

.3455E+02	.6876E-01	.3396E-01	.0000E+00	.0000E+00
.3745E+02	.6409E-01	.3198E-01	.0000E+00	.0000E+00
.4063E+02	.5934E-01	.3023E-01	.0000E+00	.0000E+00
.4411E+02	.5459E-01	.2864E-01	.0000E+00	.0000E+00
.4794E+02	.4992E-01	.2719E-01	.0000E+00	.0000E+00
.5213E+02	.4537E-01	.2582E-01	.0000E+00	.0000E+00
.5673E+02	.4101E-01	.2449E-01	.0000E+00	.0000E+00
.6177E+02	.3688E-01	.2317E-01	.0000E+00	.0000E+00
.6730E+02	.3301E-01	.2182E-01	.0000E+00	.0000E+00
.7336E+02	.2942E-01	.2044E-01	.0000E+00	.0000E+00
.8000E+02	.2612E-01	.1902E-01	.0000E+00	.0000E+00
.8729E+02	.2311E-01	.1759E-01	.0000E+00	.0000E+00
.9527E+02	.2038E-01	.1614E-01	.0000E+00	.0000E+00
.1040E+03	.1792E-01	.1471E-01	.0000E+00	.0000E+00
.1136E+03	.1572E-01	.1332E-01	.0000E+00	.0000E+00
.1242E+03	.1375E-01	.1198E-01	.0000E+00	.0000E+00
.1357E+03	.1201E-01	.1071E-01	.0000E+00	.0000E+00
.1484E+03	.1046E-01	.9518E-02	.0000E+00	.0000E+00
.1623E+03	.9101E-02	.8420E-02	.0000E+00	.0000E+00
.1775E+03	.7904E-02	.7416E-02	.0000E+00	.0000E+00
.1942E+03	.6854E-02	.6507E-02	.0000E+00	.0000E+00
.2125E+03	.5937E-02	.5691E-02	.0000E+00	.0000E+00
.2325E+03	.5138E-02	.4963E-02	.0000E+00	.0000E+00
.2545E+03	.4443E-02	.4320E-02	.0000E+00	.0000E+00
.2787E+03	.3840E-02	.3753E-02	.0000E+00	.0000E+00
.3051E+03	.3319E-02	.3258E-02	.0000E+00	.0000E+00
.3341E+03	.2869E-02	.2826E-02	.0000E+00	.0000E+00
.3659E+03	.2481E-02	.2451E-02	.0000E+00	.0000E+00
.4008E+03	.2147E-02	.2126E-02	.0000E+00	.0000E+00
.4390E+03	.1861E-02	.1845E-02	.0000E+00	.0000E+00
.4810E+03	.1614E-02	.1603E-02	.0000E+00	.0000E+00
.5269E+03	.1402E-02	.1394E-02	.0000E+00	.0000E+00
.5773E+03	.1220E-02	.1214E-02	.0000E+00	.0000E+00
.6326E+03	.1063E-02	.1059E-02	.0000E+00	.0000E+00
.6932E+03	.9280E-03	.9248E-03	.0000E+00	.0000E+00
.7597E+03	.8114E-03	.8090E-03	.0000E+00	.0000E+00
.8325E+03	.7107E-03	.7089E-03	.0000E+00	.0000E+00
.9124E+03	.6234E-03	.6220E-03	.0000E+00	.0000E+00
.1000E+04	.5477E-03	.5467E-03	.0000E+00	.0000E+00

3. Scenario 2: DEGADIS

Stability Class = F
Wind Speed $u = 2$ m/s
Roughness Length $z_0 = 0.3$ m
Source = Pipe Rupture
Averaging time $t_a = 15$ s

0   TITLE BLOCK

PIPE   SC=F   WS=2   ZO=0.3   TIME=15 MIN

        Wind velocity at reference height   2.00   m/s
        Reference height   10.00   m
0   Surface roughness length   0.300   m
0   Pasquill Stability class   F
0   Monin-Obukhov length   21.1   m
        Gaussian distribution constants   Delta   0.06940   m
                                           Beta   0.90000
0   Wind velocity power law constant   Alpha   0.53216
        Friction velocity   0.12155   m/s
0   Ambient Temperature   298.15   K
        Ambient Pressure   1.000   atm
        Ambient Absolute Humidity   1.629E-02   kg/kg BDA
        Ambient Relative Humidity   80.00   %

        Input: Mole fraction   CONCENTRATION OF C   GAS DENSITY
                                        kg/m**3   kg/m**3
                          0.00000   0.00000   1.17366
                          1.00000   3.11000   3.11000

0   Specified Gas Properties:

           Molecular weight: 76.040
           Storage temperature: 298.00   K
           Density at storage temperature and ambient pressure: 3.1100   kg/m**3
           Mean heat capacity constant: ********   [NOT USED]
           Mean heat capacity power: ********   [NOT USED]
           Upper mole fraction contour: 0.15000
           Lower mole fraction contour: 1.00000E-04
           Height for isopleths: 0.00000E-01m

    Source input data points

           Initial mass in cloud: 0.00000E-01

82

TIME	SOURCE STRENGTH	SOURCE RADIUS
s	kg/s	m
0.00000E-01	5.4200	1.0000
6023.0	5.4200	1.0000
6024.0	0.00000E-01	0.00000E-01
6025.0	0.00000E-01	0.00000E-01

0   Calculation procedure for ALPHA: 1
0   Entrainment prescription for PHI: 3
0   Layer thickness ratio used for average depth: 2.1500
0   Air entrainment coefficient used: 0.590
0   Gravity slumping velocity coefficient used: 1.150
0   Isothermal calculation
0   Heat transfer not included
0   Water transfer not included

                    *****   CALCULATED SOURCE PARAMETERS   *****

Time	Gas Radius	Height	Qstar	SZ(x=L/2.)	Mole frac C	Density	Rich No.
sec	m	m	kg/m**2/s	m		kg/m**3	
0.000000E-01	1.00000	1.100000E-05	3.001406E-02	0.131405	1.00000	3.11000	0.756144
2.000000E-02	1.00633	1.083992E-02	2.981788E-02	0.131896	0.994054	3.09849	0.756144
4.000000E-02	1.01797	2.140142E-02	2.946500E-02	0.132781	0.983515	3.07808	0.756144
6.000000E-02	1.03284	3.159621E-02	2.902608E-02	0.133904	0.970424	3.05273	0.756144
8.000000E-02	1.05018	4.136884E-02	2.853077E-02	0.135199	0.955677	3.02418	0.756144
0.120000	1.09047	5.954236E-02	2.744306E-02	0.138155	0.923390	2.96166	0.756144
0.160000	1.13633	7.585247E-02	2.630269E-02	0.141433	0.889682	2.89639	0.756144
0.320000	1.35046	0.124721	2.203915E-02	0.155660	0.764917	2.65480	0.756144
0.480000	1.58514	0.154872	1.873094E-02	0.169548	0.669396	2.46984	0.756144
0.640000	1.82420	0.173979	1.626130E-02	0.182244	0.598718	2.33298	0.756144
0.960000	2.29595	0.194866	1.293218E-02	0.204019	0.504093	2.14976	0.756144
1.28000	2.75107	0.204471	1.082327E-02	0.221949	0.444298	2.03397	0.756144
1.92000	3.61074	0.210410	8.309811E-03	0.250084	0.372881	1.89569	0.756144
2.56000	4.41450	0.209649	6.852702E-03	0.271674	0.331143	1.81487	0.756144
3.20000	5.17447	0.206517	5.892943E-03	0.289113	0.303433	1.76121	0.756144
3.84000	5.89930	0.202471	5.208040E-03	0.303690	0.283533	1.72268	0.756144
4.48000	6.59507	0.198094	4.691834E-03	0.316168	0.268467	1.69350	0.756144
5.12000	7.26624	0.193645	4.287082E-03	0.327040	0.256626	1.67057	0.756144
6.40000	8.54748	0.184946	3.689673E-03	0.345204	0.239157	1.63675	0.756144
7.68000	9.76213	0.176757	3.266787E-03	0.359889	0.226880	1.61298	0.756144
8.96000	10.9227	0.169132	2.949616E-03	0.372065	0.217813	1.59542	0.756144
10.2400	12.0378	0.162039	2.701739E-03	0.382335	0.210891	1.58202	0.756144
12.8000	14.1559	0.149241	2.337073E-03	0.398655	0.201215	1.56328	0.756144
15.3600	16.1518	0.137969	2.079820E-03	0.410886	0.195068	1.55138	0.756144
17.9200	18.0480	0.127912	1.887446E-03	0.420164	0.191155	1.54380	0.756144
20.4800	19.8601	0.118843	1.737547E-03	0.427191	0.188782	1.53921	0.756144
25.6000	23.2740	0.103031	1.518077E-03	0.436223	0.187191	1.53613	0.756144
30.7200	26.4551	8.961000E-02	1.364479E-03	0.440327	0.188524	1.53871	0.756144

83

35.8400	29.4419	7.801447E-02	1.250761E-03	0.440804	0.192005	1.54545	0.756144
46.0800	34.9290	5.892649E-02	1.094031E-03	0.433812	0.204188	1.56904	0.756144
51.2000	37.0422	5.214829E-02	1.048219E-03	0.426791	0.212728	1.58557	0.756144
56.3200	37.8519	4.933568E-02	1.036413E-03	0.418514	0.221478	1.60252	0.756144
61.4400	37.8311	4.883084E-02	1.043502E-03	0.410498	0.229575	1.61819	0.756144
66.5600	38.4857	4.680788E-02	1.033745E-03	0.404429	0.236705	1.63200	0.756144
71.6800	38.4667	4.640749E-02	1.039315E-03	0.398235	0.243556	1.64527	0.756144
76.8000	39.1189	4.455414E-02	1.028717E-03	0.393574	0.249623	1.65701	0.756144
97.2800	39.4606	4.248744E-02	1.034935E-03	0.377195	0.270377	1.69720	0.756144
102.400	40.1142	4.091198E-02	1.022846E-03	0.374738	0.274379	1.70495	0.756144
122.880	40.0132	4.016010E-02	1.033696E-03	0.364484	0.288605	1.73250	0.756144
128.000	40.6679	3.874183E-02	1.020790E-03	0.363185	0.291256	1.73763	0.756144
168.960	40.4706	3.782776E-02	1.034734E-03	0.351812	0.308481	1.77098	0.756144
184.320	41.0140	3.630512E-02	1.025522E-03	0.349067	0.313583	1.78086	0.756144
220.160	40.7750	3.573762E-02	1.035509E-03	0.343900	0.322743	1.79860	0.756144

0Source strength [kg/s] :   5.4200   Equivalent Primary source radius [m] :   1.0000
  Equivalent Primary source length [m] :   1.7725   Equivalent Primary source width [m] :   1.7725

Secondary source concentration [kg/m**3] :   1.0040   Secondary source SZ [m] :   0.34390

Contaminant flux rate: 1.03768E-03

Secondary source mass fractions... contaminant: 0.558201   air: 0.43472
            Enthalpy: 0.00000E-01   Density: 1.7986
Secondary source length [m] :   72.272   Secondary source half-width [m] :   36.136

0 Distance	Mole Fraction	Concentration	Density	Temperature	Half Width	Sz	Sy	Width at z=	0.00 m to:
								1.000E-02mole%	15.0 mole
(m)		(kg/m**3)	(kg/m**3)	(K)	(m)	(m)	(m)	(m)	(m)
36.1	0.323	1.00	1.80	298.	36.1	0.344	3.000E-02	36.2	36.2
36.7	0.309	0.961	1.77	298.	38.4	0.332	1.52	42.7	39.7
37.3	0.295	0.917	1.74	298.	41.3	0.323	2.19	47.5	43.1
37.7	0.286	0.889	1.73	298.	43.3	0.319	2.57	50.6	45.4
39.3	0.245	0.764	1.65	298.	51.1	0.314	3.80	61.7	53.7
40.9	0.208	0.649	1.58	298.	58.3	0.320	4.85	71.7	61.1
42.5	0.172	0.536	1.51	298.	64.9	0.336	5.80	80.7	67.1
45.7	0.117	0.364	1.40	298.	76.2	0.388	7.53	96.3	
48.9	8.067E-02	0.251	1.33	298.	85.4	0.457	9.10	109.	
52.1	5.734E-02	0.178	1.28	298.	93.0	0.539	10.6	120.	
55.3	4.220E-02	0.131	1.26	298.	99.2	0.628	11.9	129.	
58.5	3.229E-02	0.100	1.24	298.	104.	0.723	13.2	136.	
61.7	2.524E-02	7.854E-02	1.22	298.	109.	0.823	14.4	143.	
68.1	1.663E-02	5.174E-02	1.21	298.	116.	1.03	16.8	154.	
74.5	1.174E-02	3.653E-02	1.20	298.	121.	1.25	18.9	162.	
80.9	8.866E-03	2.759E-02	1.19	298.	119.	1.50	20.8	163.	
87.3	7.041E-03	2.191E-02	1.19	298.	118.	1.74	22.6	164.	
100.	4.897E-03	1.524E-02	1.18	298.	115.	2.21	25.8	166.	

113.	3.692E-03	1.149E-02	1.18	298.	112.	2.66	28.7	167.
126.	2.933E-03	9.127E-03	1.18	298.	110.	3.09	31.3	168.
139.	2.416E-03	7.519E-03	1.18	298.	108.	3.50	33.6	168.
151.	2.044E-03	6.361E-03	1.18	298.	106.	3.91	35.9	168.
164.	1.765E-03	5.492E-03	1.18	298.	104.	4.30	38.0	169.
177.	1.548E-03	4.817E-03	1.18	298.	102.	4.68	40.0	169.
190.	1.376E-03	4.280E-03	1.18	298.	101.	5.06	41.9	168.
203.	1.235E-03	3.844E-03	1.18	298.	99.1	5.43	43.7	168.
215.	1.119E-03	3.483E-03	1.18	298.	97.6	5.79	45.4	168.
228.	1.022E-03	3.180E-03	1.18	298.	96.1	6.14	47.1	168.
241.	9.391E-04	2.922E-03	1.18	298.	94.6	6.49	48.7	168.
254.	8.678E-04	2.700E-03	1.18	298.	93.3	6.83	50.3	167.
267.	8.060E-04	2.508E-03	1.18	298.	91.9	7.17	51.8	167.
279.	7.518E-04	2.340E-03	1.18	298.	90.6	7.50	53.3	166.
292.	7.041E-04	2.191E-03	1.18	298.	89.3	7.83	54.7	166.
305.	6.617E-04	2.059E-03	1.17	298.	88.1	8.16	56.1	165.
318.	6.238E-04	1.941E-03	1.17	298.	86.9	8.48	57.5	165.
331.	5.897E-04	1.835E-03	1.17	298.	85.7	8.79	58.8	164.
356.	5.311E-04	1.653E-03	1.17	298.	83.4	9.41	61.4	163.
369.	5.057E-04	1.574E-03	1.17	298.	82.3	9.72	62.6	162.
395.	4.612E-04	1.435E-03	1.17	298.	80.1	10.3	65.1	161.
407.	4.415E-04	1.374E-03	1.17	298.	79.1	10.6	66.2	160.
433.	4.067E-04	1.265E-03	1.17	298.	77.1	11.2	68.6	158.
446.	3.911E-04	1.217E-03	1.17	298.	76.1	11.5	69.7	157.
471.	3.631E-04	1.130E-03	1.17	298.	74.1	12.1	71.9	156.
484.	3.505E-04	1.091E-03	1.17	298.	73.2	12.3	72.9	155.
510.	3.276E-04	1.019E-03	1.17	298.	71.3	12.9	75.0	153.
523.	3.171E-04	9.869E-04	1.17	298.	70.4	13.2	76.1	152.
548.	2.981E-04	9.275E-04	1.17	298.	68.6	13.7	78.1	150.
561.	2.893E-04	9.002E-04	1.17	298.	67.7	14.0	79.1	149.
587.	2.732E-04	8.500E-04	1.17	298.	66.0	14.5	81.0	147.
612.	2.587E-04	8.049E-04	1.17	298.	64.4	15.1	82.9	145.
638.	2.455E-04	7.640E-04	1.17	298.	62.7	15.6	84.7	143.
663.	2.336E-04	7.269E-04	1.17	298.	61.1	16.1	86.5	141.
689.	2.227E-04	6.931E-04	1.17	298.	59.5	16.6	88.3	139.
715.	2.128E-04	6.621E-04	1.17	298.	58.0	17.1	90.1	136.
740.	2.036E-04	6.337E-04	1.17	298.	56.5	17.6	91.8	134.
766.	1.952E-04	6.075E-04	1.17	298.	55.0	18.1	93.4	131.
791.	1.874E-04	5.832E-04	1.17	298.	53.5	18.6	95.1	129.
817.	1.802E-04	5.607E-04	1.17	298.	52.1	19.1	96.7	126.
855.	1.703E-04	5.300E-04	1.17	298.	50.0	19.8	99.1	122.
894.	1.614E-04	5.022E-04	1.17	298.	47.9	20.5	101.	118.
932.	1.533E-04	4.771E-04	1.17	298.	45.9	21.2	104.	114.
971.	1.460E-04	4.543E-04	1.17	298.	44.0	21.9	106.	109.
1.009E+03	1.393E-04	4.335E-04	1.17	298.	42.0	22.5	108.	104.

1.047E+03	1.332E-04	4.144E-04	1.17	298.	40.2	23.2	110.	99.2
1.086E+03	1.275E-04	3.969E-04	1.17	298.	38.3	23.9	112.	93.7
1.124E+03	1.223E-04	3.807E-04	1.17	298.	36.5	24.5	114.	87.9
1.163E+03	1.175E-04	3.657E-04	1.17	298.	34.7	25.2	116.	81.5
1.201E+03	1.131E-04	3.518E-04	1.17	298.	32.9	25.8	118.	74.5
1.239E+03	1.089E-04	3.389E-04	1.17	298.	31.2	26.5	120.	66.5
1.278E+03	1.051E-04	3.269E-04	1.17	298.	29.5	27.1	122.	56.8
1.303E+03	1.026E-04	3.193E-04	1.17	298.	28.4	27.5	123.	48.5
1.329E+03	1.003E-04	3.121E-04	1.17	298.	27.3	27.9	125.	34.7
1.355E+03	9.806E-05	3.052E-04	1.17	298.	26.2	28.4	126.	
1.380E+03	9.593E-05	2.985E-04	1.17	298.	25.1	28.8	127.	
1.406E+03	9.388E-05	2.921E-04	1.17	298.	24.0	29.2	128.	
1.431E+03	9.191E-05	2.860E-04	1.17	298.	23.0	29.6	130.	
1.457E+03	9.002E-05	2.801E-04	1.17	298.	21.9	30.0	131.	
1.483E+03	8.820E-05	2.745E-04	1.17	298.	20.9	30.4	132.	
1.508E+03	8.645E-05	2.690E-04	1.17	298.	19.8	30.8	133.	
1.534E+03	8.477E-05	2.638E-04	1.17	298.	18.8	31.2	134.	
1.559E+03	8.315E-05	2.587E-04	1.17	298.	17.8	31.6	135.	
1.585E+03	8.159E-05	2.539E-04	1.17	298.	16.8	32.0	137.	
1.598E+03	8.083E-05	2.515E-04	1.17	298.	16.3	32.2	137.	
1.623E+03	7.934E-05	2.469E-04	1.17	298.	15.3	32.6	138.	
1.636E+03	7.862E-05	2.447E-04	1.17	298.	14.8	32.8	139.	
1.662E+03	7.722E-05	2.403E-04	1.17	298.	13.8	33.1	140.	
1.675E+03	7.653E-05	2.382E-04	1.17	298.	13.3	33.3	141.	
1.700E+03	7.520E-05	2.340E-04	1.17	298.	12.3	33.7	142.	
1.713E+03	7.455E-05	2.320E-04	1.17	298.	11.8	33.9	142.	
1.739E+03	7.328E-05	2.280E-04	1.17	298.	10.9	34.3	143.	
1.751E+03	7.266E-05	2.261E-04	1.17	298.	10.4	34.5	144.	
1.777E+03	7.145E-05	2.223E-04	1.17	298.	9.44	34.9	145.	
1.790E+03	7.086E-05	2.205E-04	1.17	298.	8.96	35.1	145.	
1.815E+03	6.970E-05	2.169E-04	1.17	298.	8.02	35.4	146.	
1.828E+03	6.914E-05	2.152E-04	1.17	298.	7.55	35.6	147.	
1.841E+03	6.859E-05	2.134E-04	1.17	298.	7.09	35.8	148.	
1.854E+03	6.804E-05	2.117E-04	1.17	298.	6.62	36.0	148.	
1.867E+03	6.750E-05	2.101E-04	1.17	298.	6.16	36.2	149.	
1.879E+03	6.697E-05	2.084E-04	1.17	298.	5.70	36.4	149.	
1.892E+03	6.645E-05	2.068E-04	1.17	298.	5.24	36.6	150.	
1.905E+03	6.594E-05	2.052E-04	1.17	298.	4.78	36.7	150.	
1.918E+03	6.543E-05	2.036E-04	1.17	298.	4.32	36.9	151.	
1.931E+03	6.493E-05	2.021E-04	1.17	298.	3.87	37.1	151.	
1.943E+03	6.444E-05	2.005E-04	1.17	298.	3.41	37.3	152.	
1.956E+03	6.396E-05	1.990E-04	1.17	298.	2.96	37.5	152.	
1.969E+03	6.348E-05	1.975E-04	1.17	298.	2.51	37.7	153.	
1.982E+03	6.301E-05	1.961E-04	1.17	298.	2.06	37.8	153.	
1.995E+03	6.254E-05	1.946E-04	1.17	298.	1.61	38.0	154.	

2.007E+03	6.209E-05	1.932E-04	1.17	298.	1.16	38.2	154.
2.020E+03	6.164E-05	1.918E-04	1.17	298.	0.720	38.4	155.
2.033E+03	6.119E-05	1.904E-04	1.17	298.	0.276	38.6	155.
2.046E+03	6.075E-05	1.890E-04	1.17	298.	0.000E-01	38.8	156.
2.097E+03	5.872E-05	1.827E-04	1.17	298.	0.000E-01	39.5	157.
2.123E+03	5.754E-05	1.790E-04	1.17	298.	0.000E-01	39.8	158.

For the UFL of   15.000   mole percent, and the LFL of   1.00000E-02 mole percent:

The mass of contaminant between the UFL and LFL is: 3313.2   kg.
The mass of contaminant above the LFL is: 3396.7   kg.

4. Scenario 4: DEGADIS

Stability Class = F
Wind Speed u = 2 m/s
Roughness Length z_0 = 0.3 m
Averaging time t_a = 1 min
Source: Flashed Vapor—No Aerosol

0   TITLE BLOCK

SC=F WS=2 ZO=.3 TIME=1 MIN   FLASHED VAPOR

         Wind velocity at reference height   2.00 m/s
         Reference height   10.00 m
0   Surface roughness length   0.300   m
0   Pasquill Stability class   F
0   Monin-Obukhov length   21.1   m
         Gaussian distribution constants   Delta   0.04040 m
                                            Beta   0.90000
0   Wind velocity power law constant   Alpha   0.53216
         Friction velocity   0.12155 m/s
0   Ambient Temperature   303.15 K
0   Surface Temperature   303.00 K
         Ambient Pressure   1.000 atm
         Ambient Absolute Humidity   2.223E-02 kg/kg BDA
         Ambient Relative Humidity   80.00 %

         Adiabatic Mixing: Mole fraction CONCENTRATION OF C   GAS DENSITY   Enthalpy   Temperature
                                               kg/m**3   kg/m**3   J/kg   K
                              0.00000   0.00000   1.15033   0.00000E-01   303.15
                              0.00816   0.02333   1.16668   -605.33   302.54
                              0.03165   0.09106   1.21414   -2270.0   300.80

                              0.08638   0.25068   1.31936   -5750.6   298.86
                              0.15637   0.45751   1.45241   -9533.9   297.40
                              0.24052   0.71142   1.61686   -13317.   295.26

                              0.34363   1.03219   1.82689   -17100.   291.88
                              0.45005   1.37833   2.05720   -20278.   287.17
                              0.55399   1.73727   2.30102   -22851.   280.95

                              0.65519   2.11381   2.56219   -24970.   273.18
                              0.75621   2.51252   2.83901   -26786.   265.01
                              0.84264   2.89056   3.10813   -28148.   256.33

                              0.95181   3.41270   3.48234   -29661.   244.71

90

| | 1.00000 | 3.67200 | 3.67200 | -30266. | 238.70 |

0   Specified Gas Properties:

Molecular weight:	70.910	
Storage temperature:	238.70	K
Density at storage temperature and ambient pressure:	3.6720	kg/m**3
Mean heat capacity constant:	484.20	
Mean heat capacity power:	1.0000	
Upper mole fraction contour:	1.00000E-05	
Lower mole fraction contour:	3.00000E-06	
Height for isopleths:	0.00000E-01m	

Source input data points

Initial mass in cloud: 0.00000E-01

TIME	SOURCE STRENGTH	SOURCE RADIUS
s	kg/s	m
0.00000E-01	0.24570	0.56000
6023.0	0.24570	0.56000
6024.0	0.00000E-01	0.00000E-01
6025.0	0.00000E-01	0.00000E-01

0   Calculation procedure for ALPHA: 1
0   Entrainment prescription for PHI: 3
0   Layer thickness ratio used for average depth: 2.1500
0   Air entrainment coefficient used: 0.590
0   Gravity slumping velocity coefficient used: 1.150
0   NON Isothermal calculation
0   Heat transfer calculated with correlation: 1
0   Water transfer not included

*****   CALCULATED SOURCE PARAMETERS   *****

Time sec	Gas Radius m	Height m	Qstar kg/m**2/s	SZ(x=L/2.) m	Mole frac C	Density kg/m**3	Temperature K	Rich No.
0.000000E-01	0.560000	1.104367E-05	4.593835E-02	0.106892	1.00000	3.65748	239.648	0.756144
2.000000E-02	0.562348	1.118148E-03	4.574303E-02	0.107158	0.997147	3.64749	239.910	0.756144
4.000000E-02	0.566649	2.205905E-03	4.539217E-02	0.107656	0.992071	3.62889	240.433	0.756144
6.000000E-02	0.572161	3.267072E-03	4.495132E-02	0.108298	0.985668	3.60517	241.119	0.756144
8.000000E-02	0.578612	4.297220E-03	4.444710E-02	0.109051	0.978312	3.57775	241.929	0.756144
0.120000	0.593711	6.249980E-03	4.330951E-02	0.110757	0.961652	3.51775	243.662	0.756144
0.160000	0.611071	8.053374E-03	4.207596E-02	0.112704	0.943413	3.45215	245.620	0.756144
0.320000	0.694123	1.378827E-02	3.706616E-02	0.121500	0.867768	3.19071	253.713	0.756144
0.480000	0.787792	1.755286E-02	3.273217E-02	0.130454	0.800568	2.97309	260.735	0.756144
0.640000	0.884892	1.990559E-02	2.923055E-02	0.138620	0.745684	2.80888	265.923	0.756144
0.960000	1.07894	2.203189E-02	2.415673E-02	0.152108	0.666867	2.59292	272.297	0.756144
1.28000	1.26679	2.237795E-02	2.077740E-02	0.162780	0.616011	2.45809	276.374	0.756144
1.92000	1.61794	2.084402E-02	1.662244E-02	0.177423	0.560083	2.31607	280.529	0.756144
2.56000	1.93810	1.827048E-02	1.418961E-02	0.185785	0.536834	2.25922	282.101	0.756144
3.20000	2.23094	1.549526E-02	1.260858E-02	0.189827	0.531928	2.24735	282.425	0.756144

3.84000	2.49935	1.280541E-02	1.151324E-02	0.190624	0.539661	2.26599	281.925	0.756144
4.16000	2.60382	1.171856E-02	1.116276E-02	0.189841	0.547482	2.28503	281.401	0.756144
5.12000	2.59985	1.144184E-02	1.126663E-02	0.185533	0.568406	2.33673	279.943	0.756144
5.76000	2.64139	1.096588E-02	1.115918E-02	0.183922	0.578210	2.36110	279.257	0.756144
7.68000	2.62620	1.063288E-02	1.130589E-02	0.179296	0.601704	2.42104	277.484	0.756144
10.2400	2.61891	1.045340E-02	1.138330E-02	0.176766	0.615082	2.45578	276.436	0.756144
14.0800	2.61687	1.038282E-02	1.141200E-02	0.175728	0.620726	2.47054	275.989	0.756144
16.6400	2.69232	9.199586E-03	1.119801E-02	0.173556	0.635904	2.51048	274.779	0.756144
19.2000	2.63951	8.822231E-03	1.145987E-02	0.169607	0.656768	2.56651	273.044	0.756144
23.0400	2.60004	8.542347E-03	1.166825E-02	0.166644	0.673657	2.61095	271.773	0.756144
25.6000	2.58769	8.455823E-03	1.173490E-02	0.165726	0.679045	2.62515	271.372	0.756144
30.7200	2.57721	8.382804E-03	1.179171E-02	0.164951	0.683610	2.63728	271.025	0.756144
40.9600	2.57317	8.354054E-03	1.181307E-02	0.164629	0.685354	2.64222	270.863	0.756144

OSource strength [kg/s] : 0.24570   Equivalent Primary source radius [m] : 0.56000

Equivalent Primary source length [m] : 0.99257   Equivalent Primary source width [m] : 0.99257

Secondary source concentration [kg/m**3] : 2.2296   Secondary source SZ [m] : 0.16463

Contaminant flux rate: 1.18118E-02

Secondary source mass fractions... contaminant: 0.843847   air: 0.15276

Enthalpy: -25540.   Density: 2.6422

Secondary source length [m] : 4.5608   Secondary source half-width [m] : 2.2804

O Distance	Mole Fraction	Concentration	Density	Gamma	Temperature	Half Width	Sz	Sy	Width at z= 0.00 m to:	
									3.000E-04mole%	1.000E-03mole%
(m)		(kg/m**3)	(kg/m**3)		(K)	(m)	(m)	(m)	(m)	(m)
2.28	0.685	2.23	2.6125	0.656	272.	2.28	0.163	3.000E-02	2.39	2.38
2.33	0.666	2.15	2.5540	0.652	275.	2.69	0.147	0.104	3.06	3.04
2.38	0.648	2.09	2.5023	0.648	276.	3.12	0.136	0.148	3.64	3.61
2.43	0.628	2.01	2.4407	0.642	279.	3.54	0.128	0.186	4.20	4.17
2.48	0.603	1.92	2.3705	0.637	281.	3.96	0.122	0.220	4.74	4.70
2.53	0.579	1.83	2.3053	0.631	283.	4.37	0.118	0.253	5.26	5.21
2.58	0.555	1.74	2.2412	0.626	285.	4.77	0.115	0.284	5.76	5.71
2.63	0.527	1.64	2.1725	0.622	288.	5.15	0.114	0.315	6.25	6.20
2.68	0.503	1.56	2.1138	0.618	289.	5.53	0.112	0.344	6.73	6.66
2.78	0.454	1.39	1.9995	0.609	293.	6.24	0.112	0.401	7.63	7.56
2.88	0.410	1.25	1.8977	0.600	296.	6.91	0.113	0.455	8.48	8.40
2.98	0.369	1.11	1.8145	0.596	298.	7.53	0.115	0.508	9.28	9.19
3.08	0.330	0.990	1.7388	0.594	299.	8.12	0.117	0.559	10.0	9.93
3.28	0.267	0.793	1.6212	0.594	301.	9.20	0.125	0.658	11.4	11.3
3.48	0.216	0.637	1.5293	0.595	301.	10.2	0.135	0.752	12.7	12.5
3.68	0.176	0.517	1.4584	0.596	302.	11.0	0.147	0.842	13.8	13.6
3.88	0.145	0.424	1.4036	0.597	302.	11.8	0.160	0.929	14.8	14.7
4.08	0.121	0.352	1.3610	0.598	302.	12.5	0.173	1.01	15.8	15.6
4.28	0.102	0.296	1.3276	0.599	302.	13.1	0.188	1.09	16.6	16.4
4.48	8.662E-02	0.251	1.3011	0.600	303.	13.7	0.203	1.17	17.4	17.2
4.68	7.451E-02	0.216	1.2799	0.600	303.	14.2	0.218	1.25	18.2	17.9

5.08	5.676E-02	0.164	1.2490	0.602	303.	15.1	0.250	1.40	19.5	19.2
5.48	4.461E-02	0.129	1.2278	0.602	303.	15.9	0.283	1.54	20.7	20.4
6.28	2.974E-02	8.554E-02	1.2020	0.604	303.	17.2	0.349	1.80	22.7	22.3
7.08	2.152E-02	6.175E-02	1.1877	0.605	303.	18.3	0.414	2.04	24.4	24.0
7.88	1.629E-02	4.668E-02	1.1786	0.605	303.	19.2	0.480	2.27	25.9	25.4
8.68	1.288E-02	3.687E-02	1.1726	0.605	303.	19.3	0.555	2.48	26.5	25.9
9.48	1.057E-02	3.026E-02	1.1686	0.605	303.	19.1	0.631	2.68	26.8	26.2
10.3	8.931E-03	2.555E-02	1.1658	0.605	303.	19.0	0.704	2.86	27.1	26.4
11.1	7.707E-03	2.204E-02	1.1637	0.605	303.	18.8	0.776	3.04	27.3	26.6
11.9	6.761E-03	1.933E-02	1.1620	0.605	303.	18.7	0.845	3.20	27.5	26.8
13.5	5.406E-03	1.545E-02	1.1597	0.605	303.	18.4	0.978	3.50	28.0	27.2
15.1	4.481E-03	1.280E-02	1.1581	0.605	303.	18.1	1.11	3.78	28.4	27.5
18.3	3.313E-03	9.465E-03	1.1561	0.606	303.	17.7	1.35	4.29	29.0	28.0
21.5	2.611E-03	7.455E-03	1.1548	0.606	303.	17.3	1.57	4.74	29.6	28.5
24.7	2.145E-03	6.124E-03	1.1540	0.606	303.	16.9	1.79	5.15	30.1	28.9
27.9	1.815E-03	5.181E-03	1.1535	0.607	303.	16.6	1.99	5.53	30.6	29.2
34.3	1.381E-03	3.941E-03	1.1527	0.607	303.	16.0	2.38	6.23	31.4	29.8
40.7	1.109E-03	3.166E-03	1.1522	0.607	303.	15.4	2.75	6.85	32.1	30.3
47.1	9.240E-04	2.638E-03	1.1519	0.607	303.	14.9	3.10	7.42	32.7	30.7
53.5	7.903E-04	2.256E-03	1.1517	0.609	303.	14.4	3.43	7.95	33.2	31.1
66.3	6.108E-04	1.743E-03	1.1514	0.609	303.	13.6	4.06	8.92	34.1	31.7
79.1	4.961E-04	1.416E-03	1.1512	0.609	303.	12.8	4.65	9.79	34.9	32.2
91.9	4.168E-04	1.190E-03	1.1511	0.609	303.	12.1	5.21	10.6	35.6	32.6
105.	3.588E-04	1.024E-03	1.1509	0.609	303.	11.4	5.74	11.3	36.2	32.9
117.	3.147E-04	8.981E-04	1.1509	0.614	303.	10.8	6.26	12.0	36.8	33.2
130.	2.800E-04	7.991E-04	1.1508	0.614	303.	10.2	6.75	12.7	37.3	33.4
143.	2.520E-04	7.193E-04	1.1508	0.614	303.	9.69	7.23	13.3	37.7	33.6
156.	2.290E-04	6.536E-04	1.1507	0.614	303.	9.16	7.70	13.9	38.1	33.8
169.	2.098E-04	5.987E-04	1.1507	0.614	303.	8.65	8.15	14.5	38.5	33.9
181.	1.934E-04	5.521E-04	1.1507	0.614	303.	8.16	8.60	15.0	38.9	34.0
194.	1.794E-04	5.120E-04	1.1506	0.621	303.	7.69	9.03	15.6	39.2	34.1
207.	1.673E-04	4.773E-04	1.1506	0.621	303.	7.23	9.45	16.1	39.5	34.2
220.	1.566E-04	4.470E-04	1.1506	0.621	303.	6.79	9.87	16.6	39.8	34.3
233.	1.472E-04	4.201E-04	1.1506	0.621	303.	6.36	10.3	17.1	40.0	34.4
245.	1.388E-04	3.963E-04	1.1506	0.621	303.	5.94	10.7	17.5	40.3	34.4
258.	1.314E-04	3.749E-04	1.1506	0.621	303.	5.54	11.1	18.0	40.5	34.4
271.	1.246E-04	3.557E-04	1.1505	0.621	303.	5.14	11.5	18.4	40.7	34.4
284.	1.185E-04	3.383E-04	1.1505	0.621	303.	4.75	11.8	18.9	41.0	34.4
297.	1.130E-04	3.225E-04	1.1505	0.621	303.	4.38	12.2	19.3	41.2	34.4
309.	1.080E-04	3.081E-04	1.1505	0.621	303.	4.01	12.6	19.7	41.3	34.4
322.	1.033E-04	2.949E-04	1.1505	0.633	303.	3.64	12.9	20.1	41.5	34.4
335.	9.908E-05	2.828E-04	1.1505	0.633	303.	3.29	13.3	20.5	41.7	34.4
348.	9.517E-05	2.716E-04	1.1505	0.633	303.	2.94	13.7	20.9	41.9	34.4
361.	9.154E-05	2.613E-04	1.1505	0.633	303.	2.60	14.0	21.3	42.0	34.3
373.	8.818E-05	2.517E-04	1.1505	0.633	303.	2.26	14.4	21.7	42.1	34.3

386.	8.505E-05	2.427E-04	1.1505	0.633	303.	1.93	14.7	22.1	42.3	34.2
399.	8.213E-05	2.344E-04	1.1505	0.633	303.	1.61	15.0	22.4	42.4	34.2
412.	7.940E-05	2.266E-04	1.1505	0.632	303.	1.29	15.4	22.8	42.5	34.1
425.	7.684E-05	2.193E-04	1.1505	0.632	303.	0.975	15.7	23.1	42.7	34.0
437.	7.444E-05	2.124E-04	1.1505	0.632	303.	0.665	16.0	23.5	42.8	34.0
450.	7.218E-05	2.060E-04	1.1505	0.632	303.	0.360	16.4	23.8	42.9	33.9
463.	7.005E-05	1.999E-04	1.1505	0.632	303.	5.967E-02	16.7	24.2	43.0	33.8
476.	6.804E-05	1.942E-04	1.1504	0.632	303.	0.000E-01	17.0	24.2	42.8	33.6
489.	6.570E-05	1.875E-04	1.1504	0.632	303.	0.000E-01	17.3	24.6	43.2	33.7
501.	6.261E-05	1.787E-04	1.1504	0.632	303.	0.000E-01	17.6	24.9	43.4	33.7
514.	6.013E-05	1.716E-04	1.1504	0.601	303.	0.000E-01	17.9	25.3	43.7	33.8
527.	5.780E-05	1.650E-04	1.1504	0.601	303.	0.000E-01	18.3	25.6	44.0	33.9
553.	5.353E-05	1.528E-04	1.1504	0.601	303.	0.000E-01	18.9	26.3	44.6	34.0
565.	5.157E-05	1.472E-04	1.1504	0.601	303.	0.000E-01	19.2	26.6	44.8	34.1
591.	4.798E-05	1.369E-04	1.1504	0.601	303.	0.000E-01	19.8	27.3	45.4	34.1
604.	4.632E-05	1.322E-04	1.1504	0.601	303.	0.000E-01	20.1	27.6	45.6	34.2
629.	4.325E-05	1.234E-04	1.1504	0.601	303.	0.000E-01	20.7	28.2	46.1	34.2
642.	4.184E-05	1.194E-04	1.1504	0.601	303.	0.000E-01	21.0	28.6	46.4	34.2
668.	3.921E-05	1.119E-04	1.1504	0.601	303.	0.000E-01	21.6	29.2	46.9	34.2
681.	3.799E-05	1.084E-04	1.1504	0.601	303.	0.000E-01	21.9	29.6	47.1	34.2
706.	3.571E-05	1.019E-04	1.1504	0.601	303.	0.000E-01	22.4	30.2	47.6	34.1
719.	3.465E-05	9.890E-05	1.1504	0.601	303.	0.000E-01	22.7	30.5	47.8	34.1
745.	3.267E-05	9.325E-05	1.1504	0.601	303.	0.000E-01	23.3	31.2	48.2	33.9
757.	3.175E-05	9.060E-05	1.1504	0.600	303.	0.000E-01	23.6	31.5	48.4	33.9
783.	3.001E-05	8.565E-05	1.1504	0.600	303.	0.000E-01	24.1	32.2	48.8	33.7
796.	2.920E-05	8.332E-05	1.1504	0.600	303.	0.000E-01	24.4	32.5	49.0	33.6
821.	2.766E-05	7.895E-05	1.1504	0.600	303.	0.000E-01	24.9	33.2	49.4	33.4
834.	2.694E-05	7.690E-05	1.1504	0.600	303.	0.000E-01	25.2	33.5	49.6	33.3
860.	2.559E-05	7.302E-05	1.1504	0.600	303.	0.000E-01	25.8	34.1	50.0	33.1
873.	2.495E-05	7.120E-05	1.1504	0.600	303.	0.000E-01	26.0	34.4	50.1	32.9
898.	2.374E-05	6.775E-05	1.1504	0.600	303.	0.000E-01	26.6	35.1	50.5	32.6
911.	2.317E-05	6.612E-05	1.1504	0.600	303.	0.000E-01	26.8	35.4	50.6	32.5
937.	2.209E-05	6.303E-05	1.1504	0.600	303.	0.000E-01	27.3	36.1	51.0	32.1
949.	2.157E-05	6.157E-05	1.1504	0.600	303.	0.000E-01	27.6	36.4	51.1	31.9
975.	2.060E-05	5.880E-05	1.1504	0.600	303.	0.000E-01	28.1	37.0	51.4	31.5
988.	2.014E-05	5.749E-05	1.1504	0.600	303.	0.000E-01	28.4	37.3	51.5	31.3
1.013E+03	1.927E-05	5.499E-05	1.1504	0.599	303.	0.000E-01	28.9	38.0	51.8	30.8
1.039E+03	1.845E-05	5.265E-05	1.1504	0.599	303.	0.000E-01	29.4	38.6	52.1	30.2
1.065E+03	1.768E-05	5.046E-05	1.1504	0.599	303.	0.000E-01	29.9	39.3	52.3	29.6
1.090E+03	1.696E-05	4.841E-05	1.1504	0.599	303.	0.000E-01	30.4	39.9	52.5	29.0
1.116E+03	1.629E-05	4.648E-05	1.1504	0.599	303.	0.000E-01	30.9	40.5	52.7	28.3
1.141E+03	1.565E-05	4.466E-05	1.1504	0.599	303.	0.000E-01	31.4	41.2	52.9	27.6
1.167E+03	1.505E-05	4.296E-05	1.1504	0.599	303.	0.000E-01	31.9	41.8	53.1	26.7
1.193E+03	1.449E-05	4.135E-05	1.1504	0.599	303.	0.000E-01	32.4	42.4	53.2	25.8
1.218E+03	1.395E-05	3.982E-05	1.1503	0.598	303.	0.000E-01	32.9	43.1	53.4	24.9

1.244E+03	1.345E-05	3.839E-05	1.1503	0.598	303.	0.000E-01	33.4	43.7	53.5	23.8
1.269E+03	1.297E-05	3.703E-05	1.1503	0.598	303.	0.000E-01	33.8	44.3	53.6	22.6
1.295E+03	1.252E-05	3.574E-05	1.1503	0.598	303.	0.000E-01	34.3	45.0	53.7	21.3
1.321E+03	1.210E-05	3.452E-05	1.1503	0.598	303.	0.000E-01	34.8	45.6	53.8	19.9
1.346E+03	1.169E-05	3.336E-05	1.1503	0.598	303.	0.000E-01	35.2	46.2	53.9	18.3
1.372E+03	1.130E-05	3.226E-05	1.1503	0.597	303.	0.000E-01	35.7	46.8	53.9	16.4
1.397E+03	1.094E-05	3.122E-05	1.1503	0.597	303.	0.000E-01	36.2	47.5	54.0	14.2
1.423E+03	1.059E-05	3.022E-05	1.1503	0.597	303.	0.000E-01	36.6	48.1	54.0	11.5
1.449E+03	1.026E-05	2.927E-05	1.1503	0.597	303.	0.000E-01	37.1	48.7	54.0	7.76
1.474E+03	9.941E-06	2.837E-05	1.1503	0.597	303.	0.000E-01	37.5	49.3	54.0	
1.500E+03	9.639E-06	2.751E-05	1.1503	0.597	303.	0.000E-01	38.0	49.9	54.0	
1.525E+03	9.351E-06	2.669E-05	1.1503	0.597	303.	0.000E-01	38.4	50.6	53.9	
1.551E+03	9.076E-06	2.590E-05	1.1503	0.596	303.	0.000E-01	38.9	51.2	53.8	
1.589E+03	8.686E-06	2.479E-05	1.1503	0.596	303.	0.000E-01	39.6	52.1	53.7	
1.628E+03	8.321E-06	2.375E-05	1.1503	0.596	303.	0.000E-01	40.2	53.0	53.6	
1.666E+03	7.978E-06	2.277E-05	1.1503	0.596	303.	0.000E-01	40.9	54.0	53.4	
1.705E+03	7.657E-06	2.185E-05	1.1503	0.595	303.	0.000E-01	41.5	54.9	53.1	
1.743E+03	7.355E-06	2.099E-05	1.1503	0.595	303.	0.000E-01	42.2	55.8	52.8	
1.781E+03	7.125E-06	2.033E-05	1.1503	0.595	303.	0.000E-01	42.8	56.7	52.7	
1.820E+03	6.854E-06	1.956E-05	1.1503	0.594	303.	0.000E-01	43.4	57.6	52.4	
1.858E+03	6.599E-06	1.883E-05	1.1503	0.594	303.	0.000E-01	44.1	58.5	52.0	
1.897E+03	6.357E-06	1.814E-05	1.1503	0.594	303.	0.000E-01	44.7	59.5	51.5	
1.935E+03	6.129E-06	1.749E-05	1.1503	0.594	303.	0.000E-01	45.3	60.4	51.0	
1.973E+03	5.913E-06	1.688E-05	1.1503	0.593	303.	0.000E-01	45.9	61.3	50.5	
2.012E+03	5.708E-06	1.629E-05	1.1503	0.593	303.	0.000E-01	46.5	62.2	49.9	
2.050E+03	5.515E-06	1.574E-05	1.1503	0.593	303.	0.000E-01	47.2	63.1	49.2	
2.089E+03	5.330E-06	1.521E-05	1.1503	0.592	303.	0.000E-01	47.8	64.0	48.5	
2.127E+03	5.156E-06	1.471E-05	1.1503	0.592	303.	0.000E-01	48.4	64.9	47.8	
2.178E+03	4.936E-06	1.409E-05	1.1503	0.591	303.	0.000E-01	49.2	66.1	46.6	
2.229E+03	4.730E-06	1.350E-05	1.1503	0.591	303.	0.000E-01	50.0	67.3	45.4	
2.281E+03	4.536E-06	1.295E-05	1.1503	0.591	303.	0.000E-01	50.7	68.5	44.1	
2.332E+03	4.355E-06	1.243E-05	1.1503	0.590	303.	0.000E-01	51.5	69.7	42.6	
2.383E+03	4.185E-06	1.194E-05	1.1503	0.590	303.	0.000E-01	52.3	70.9	40.9	
2.434E+03	4.024E-06	1.148E-05	1.1503	0.589	303.	0.000E-01	53.1	72.1	39.1	
2.485E+03	3.873E-06	1.105E-05	1.1503	0.589	303.	0.000E-01	53.8	73.3	37.0	
2.537E+03	3.730E-06	1.064E-05	1.1503	0.588	303.	0.000E-01	54.6	74.5	34.7	
2.588E+03	3.595E-06	1.026E-05	1.1503	0.588	303.	0.000E-01	55.3	75.7	32.2	
2.639E+03	3.467E-06	9.895E-06	1.1503	0.587	303.	0.000E-01	56.1	76.8	29.2	
2.690E+03	3.346E-06	9.550E-06	1.1503	0.586	303.	0.000E-01	56.8	78.0	25.8	
2.754E+03	3.204E-06	9.144E-06	1.1503	0.586	303.	0.000E-01	57.7	79.5	20.4	
2.818E+03	3.071E-06	8.763E-06	1.1503	0.585	303.	0.000E-01	58.7	81.0	12.3	
2.882E+03	2.946E-06	8.406E-06	1.1503	0.584	303.	0.000E-01	59.6	82.4		
2.946E+03	2.828E-06	8.071E-06	1.1503	0.584	303.	0.000E-01	60.5	83.9		
3.010E+03	2.718E-06	7.756E-06	1.1503	0.583	303.	0.000E-01	61.4	85.4		
3.074E+03	2.614E-06	7.459E-06	1.1503	0.582	303.	0.000E-01	62.2	86.8		

3.138E+03	2.516E-06	7.179E-06	1.1503	0.581	303.	0.000E-01	63.1	88.3
3.202E+03	2.423E-06	6.915E-06	1.1503	0.580	303.	0.000E-01	64.0	89.7
3.266E+03	2.336E-06	6.666E-06	1.1503	0.580	303.	0.000E-01	64.9	91.2
3.343E+03	2.237E-06	6.384E-06	1.1503	0.579	303.	0.000E-01	65.9	92.9
3.420E+03	2.144E-06	6.120E-06	1.1503	0.578	303.	0.000E-01	66.9	94.7
3.497E+03	2.058E-06	5.872E-06	1.1503	0.577	303.	0.000E-01	67.9	96.4
3.573E+03	1.976E-06	5.639E-06	1.1503	0.576	303.	0.000E-01	69.0	98.1
3.650E+03	1.899E-06	5.420E-06	1.1503	0.574	303.	0.000E-01	70.0	99.8
3.727E+03	1.827E-06	5.214E-06	1.1503	0.573	303.	0.000E-01	70.9	102.
3.804E+03	1.759E-06	5.019E-06	1.1503	0.572	303.	0.000E-01	71.9	103.
3.881E+03	1.694E-06	4.836E-06	1.1503	0.571	303.	0.000E-01	72.9	105.
3.957E+03	1.634E-06	4.662E-06	1.1503	0.570	303.	0.000E-01	73.9	107.
4.034E+03	1.576E-06	4.498E-06	1.1503	0.569	303.	0.000E-01	74.9	108.
4.111E+03	1.521E-06	4.342E-06	1.1503	0.568	303.	0.000E-01	75.8	110.
4.137E+03	1.504E-06	4.292E-06	1.1503	0.567	303.	0.000E-01	76.1	111.

For the UFL of   1.00000E-03 mole percent, and the LFL of   3.00000E-04 mole percent:

The mass of contaminant between the UFL and LFL is: -74.877   kg.
The mass of contaminant above the LFL is: 142.52   kg.

5. Scenario 4: DEGADIS

Stability Class = F
Wind Speed u = 2 m/s
Roughness Length z_0 = 0.3 m
Averaging time t_a = 1 min
Source: Flashed Vapor and Aerosol

0   TITLE BLOCK

SC=F WS=2 ZO=.3 TIME=1 MIN   AEROSOL

          Wind velocity at reference height   2.00 m/s
          Reference height   10.00 m
0   Surface roughness length   0.300 m
0   Pasquill Stability class   F
0   Monin-Obukhov length   21.1   m
          Gaussian distribution constants   Delta   0.04040 m
                                            Beta   0.90000
0   Wind velocity power law constant Alpha   0.53216
          Friction velocity   0.12155 m/s
0   Ambient Temperature   303.15 K
          Ambient Pressure   1.000 atm
          Ambient Absolute Humidity   2.223E-02 kg/kg BDA
          Ambient Relative Humidity   80.00 %

          Input: Mole fraction   CONCENTRATION OF C   GAS DENSITY
                                            kg/m**3   kg/m**3
                          0.00000   0.00000   1.15033
                          1.00000   8.18800   8.18800

0   Specified Gas Properties:

          Molecular weight: 203.60
          Storage temperature: 303.00   K
          Density at storage temperature and ambient pressure: 8.1880   kg/m**3
          Mean heat capacity constant: 484.20
          Mean heat capacity power: 1.0000
          Upper mole fraction contour: 1.00000E-05
          Lower mole fraction contour: 1.05000E-06
          Height for isopleths: 0.00000E-01m

Source input data points

          Initial mass in cloud: 0.00000E-01

	TIME	SOURCE STRENGTH	SOURCE RADIUS
	s	kg/s	m
	0.00000E-01	0.70800	0.56000
	6023.0	0.70800	0.56000
	6024.0	0.00000E-01	0.00000E-01
	6025.0	0.00000E-01	0.00000E-01

0   Calculation procedure for ALPHA: 1
0   Entrainment prescription for PHI: 3
0   Layer thickness ratio used for average depth: 2.1500
0   Air entrainment coefficient used: 0.590
0   Gravity slumping velocity coefficient used: 1.150
0   Isothermal calculation
0   Heat transfer not included
0   Water transfer not included

                    *****   CALCULATED SOURCE PARAMETERS   *****

Time	Gas Radius	Height	Qstar	SZ(x=L/2.)	Mole frac C	Density	Rich No.
sec	m	m	kg/m**2/s	m		kg/m**3	
0.000000E-01	0.560000	1.100000E-05	4.507848E-02	6.239602E-02	1.00000	8.18800	0.756144
1.000000E-02	0.561647	8.310223E-04	4.494259E-02	6.250797E-02	0.997175	8.16812	0.756144
3.000000E-02	0.568753	2.436671E-03	4.437064E-02	6.296662E-02	0.985834	8.08831	0.756144
4.000000E-02	0.573444	3.215570E-03	4.400103E-02	6.326735E-02	0.978516	8.03680	0.756144
6.000000E-02	0.584488	4.715764E-03	4.315505E-02	6.396881E-02	0.961798	7.91914	0.756144
8.000000E-02	0.597295	6.132056E-03	4.221448E-02	6.477088E-02	0.943265	7.78872	0.756144
0.120000	0.626801	8.700485E-03	4.019795E-02	6.657456E-02	0.903730	7.51049	0.756144
0.160000	0.660007	1.092442E-02	3.815037E-02	6.853431E-02	0.863864	7.22992	0.756144
0.320000	0.811137	1.704903E-02	3.100055E-02	7.664051E-02	0.726884	6.26590	0.756144
0.480000	0.971907	2.032872E-02	2.588767E-02	8.407457E-02	0.631115	5.59191	0.756144
0.640000	1.13247	2.211498E-02	2.225809E-02	9.055291E-02	0.564291	5.12162	0.756144
0.960000	1.44344	2.356190E-02	1.755987E-02	0.101075	0.479455	4.52458	0.756144
1.28000	1.73860	2.374677E-02	1.467413E-02	0.109222	0.428541	4.16626	0.756144
1.92000	2.28737	2.281931E-02	1.130555E-02	0.121013	0.371228	3.76291	0.756144
2.56000	2.79221	2.137076E-02	9.383096E-03	0.129088	0.340657	3.54776	0.756144
3.84000	3.70614	1.832915E-02	7.241489E-03	0.138967	0.311671	3.34377	0.756144
5.12000	4.52603	1.555928E-02	6.060950E-03	0.143956	0.301809	3.27436	0.756144
7.68000	5.96448	1.101006E-02	4.786059E-03	0.145889	0.307718	3.31595	0.756144
10.2400	7.00486	7.971871E-03	4.214909E-03	0.141146	0.334802	3.50656	0.756144
11.5200	7.41361	6.828644E-03	4.042322E-03	0.137769	0.352681	3.63238	0.756144
14.0800	7.35632	6.340797E-03	4.127799E-03	0.130546	0.388096	3.88162	0.756144
17.9200	7.29580	5.843192E-03	4.220100E-03	0.123180	0.430142	4.17753	0.756144
20.4800	7.26667	5.618604E-03	4.264115E-03	0.119860	0.451405	4.32717	0.756144
25.6000	7.22532	5.330109E-03	4.323988E-03	0.115615	0.480996	4.53542	0.756144
30.7200	7.19828	5.169192E-03	4.360001E-03	0.113274	0.498570	4.65910	0.756144
40.9600	7.16739	5.022992E-03	4.396106E-03	0.111201	0.514916	4.77414	0.756144
56.3200	7.14792	4.956103E-03	4.415201E-03	0.110231	0.522203	4.82542	0.756144

0Source strength [kg/s] :   0.70800   Equivalent Primary source radius [m] :   0.56000

99

Equivalent Primary source length [m] :   0.99257   Equivalent Primary source width [m] :   0.99257

Secondary source concentration [kg/m**3] :   4.2762   Secondary source SZ [m] :   0.11023

Contaminant flux rate: 4.41087E-03

Secondary source mass fractions... contaminant: 0.886176   air: 0.11135
        Enthalpy: 0.00000E-01   Density: 4.8254
Secondary source length [m] :   12.669   Secondary source half-width [m] :   6.3347

O Distance (m)	Mole Fraction	Concentration (kg/m**3)	Density (kg/m**3)	Temperature (K)	Half Width (m)	Sz (m)	Sy (m)	Width at z= 0.00 m to: 1.050E-04mole% (m)	1.000E-03mole% (m)
6.33	0.522	4.28	4.83	303.	6.33	0.110	3.000E-02	6.44	6.43
6.36	0.514	4.21	4.77	303.	6.65	0.107	0.112	7.05	7.01
6.41	0.508	4.16	4.73	303.	7.34	0.100	0.191	8.03	7.97
6.43	0.505	4.14	4.71	303.	7.69	9.732E-02	0.222	8.49	8.42
6.53	0.488	4.00	4.59	303.	9.11	8.869E-02	0.325	10.3	10.2
6.63	0.470	3.85	4.46	303.	10.5	8.258E-02	0.414	12.0	11.9
6.73	0.450	3.69	4.32	303.	11.9	7.834E-02	0.495	13.7	13.5
6.93	0.400	3.28	3.97	303.	14.6	7.401E-02	0.645	16.9	16.7
7.13	0.345	2.83	3.58	303.	17.1	7.335E-02	0.786	19.9	19.6
7.53	0.249	2.04	2.90	303.	21.6	7.833E-02	1.05	25.3	24.9
7.93	0.173	1.41	2.37	303.	25.3	8.969E-02	1.29	29.8	29.3
8.73	8.650E-02	0.709	1.76	303.	30.9	0.123	1.74	36.7	36.1
9.53	4.877E-02	0.400	1.49	303.	34.8	0.165	2.14	41.8	41.0
11.1	2.101E-02	0.172	1.30	303.	40.1	0.259	2.85	49.1	48.0
12.7	1.157E-02	9.482E-02	1.23	303.	43.7	0.359	3.47	54.3	52.9
15.9	5.173E-03	4.239E-02	1.19	303.	48.4	0.564	4.54	61.7	59.8
19.1	3.027E-03	2.480E-02	1.17	303.	51.0	0.770	5.47	66.5	64.1
22.3	2.055E-03	1.684E-02	1.16	303.	50.3	0.989	6.28	67.6	64.8
25.5	1.534E-03	1.257E-02	1.16	303.	49.7	1.20	7.00	68.6	65.4
28.7	1.213E-03	9.940E-03	1.16	303.	49.1	1.39	7.65	69.4	65.9
31.9	9.974E-04	8.173E-03	1.16	303.	48.6	1.58	8.25	70.2	66.3
38.3	7.288E-04	5.972E-03	1.16	303.	47.6	1.94	9.33	71.5	67.0
44.7	5.692E-04	4.664E-03	1.15	303.	46.8	2.28	10.3	72.6	67.5
51.1	4.644E-04	3.806E-03	1.15	303.	46.0	2.61	11.2	73.6	67.9
57.5	3.908E-04	3.202E-03	1.15	303.	45.3	2.92	12.0	74.5	68.3
70.3	2.948E-04	2.416E-03	1.15	303.	43.9	3.51	13.5	76.0	68.8
83.1	2.353E-04	1.928E-03	1.15	303.	42.7	4.06	14.9	77.3	69.1
95.9	1.951E-04	1.598E-03	1.15	303.	41.6	4.59	16.1	78.4	69.4
109.	1.662E-04	1.362E-03	1.15	303.	40.6	5.10	17.2	79.4	69.5
122.	1.444E-04	1.184E-03	1.15	303.	39.7	5.59	18.3	80.3	69.6
134.	1.276E-04	1.045E-03	1.15	303.	38.8	6.06	19.3	81.1	69.6
147.	1.141E-04	9.349E-04	1.15	303.	37.9	6.52	20.3	81.8	69.6
160.	1.031E-04	8.448E-04	1.15	303.	37.1	6.96	21.2	82.5	69.5
173.	9.397E-05	7.700E-04	1.15	303.	36.3	7.40	22.1	83.1	69.4

186.	8.627E-05	7.069E-04	1.15	303.	35.6	7.82	22.9	83.7	69.2
198.	7.970E-05	6.531E-04	1.15	303.	34.9	8.24	23.7	84.3	69.1
211.	7.403E-05	6.066E-04	1.15	303.	34.2	8.64	24.5	84.8	68.9
224.	6.909E-05	5.661E-04	1.15	303.	33.5	9.04	25.3	85.2	68.7
237.	6.474E-05	5.305E-04	1.15	303.	32.8	9.43	26.0	85.7	68.4
250.	6.089E-05	4.990E-04	1.15	303.	32.2	9.82	26.8	86.1	68.2
262.	5.746E-05	4.709E-04	1.15	303.	31.6	10.2	27.5	86.5	67.9
288.	5.161E-05	4.229E-04	1.15	303.	30.4	10.9	28.8	87.2	67.3
301.	4.910E-05	4.023E-04	1.15	303.	29.8	11.3	29.5	87.6	67.0
326.	4.472E-05	3.665E-04	1.15	303.	28.7	12.0	30.7	88.2	66.3
339.	4.281E-05	3.508E-04	1.15	303.	28.1	12.4	31.4	88.5	65.9
365.	3.941E-05	3.230E-04	1.15	303.	27.1	13.0	32.5	89.0	65.2
378.	3.791E-05	3.106E-04	1.15	303.	26.5	13.4	33.1	89.3	64.8
403.	3.520E-05	2.885E-04	1.15	303.	25.5	14.0	34.3	89.7	64.0
416.	3.399E-05	2.785E-04	1.15	303.	25.1	14.4	34.8	90.0	63.6
442.	3.178E-05	2.604E-04	1.15	303.	24.1	15.0	35.9	90.4	62.7
454.	3.078E-05	2.522E-04	1.15	303.	23.6	15.3	36.4	90.6	62.3
480.	2.896E-05	2.373E-04	1.15	303.	22.7	15.9	37.4	90.9	61.3
506.	2.733E-05	2.239E-04	1.15	303.	21.8	16.6	38.5	91.2	60.4
531.	2.587E-05	2.120E-04	1.15	303.	21.0	17.2	39.4	91.5	59.4
557.	2.455E-05	2.012E-04	1.15	303.	20.1	17.8	40.4	91.8	58.4
582.	2.336E-05	1.914E-04	1.15	303.	19.3	18.3	41.3	92.1	57.4
608.	2.228E-05	1.825E-04	1.15	303.	18.5	18.9	42.2	92.3	56.3
634.	2.129E-05	1.744E-04	1.15	303.	17.7	19.5	43.1	92.5	55.2
659.	2.038E-05	1.670E-04	1.15	303.	16.9	20.1	44.0	92.7	54.1
685.	1.954E-05	1.601E-04	1.15	303.	16.1	20.6	44.9	92.9	52.9
710.	1.877E-05	1.538E-04	1.15	303.	15.4	21.2	45.7	93.0	51.7
736.	1.806E-05	1.480E-04	1.15	303.	14.7	21.7	46.5	93.2	50.5
762.	1.740E-05	1.425E-04	1.15	303.	13.9	22.2	47.3	93.3	49.2
787.	1.678E-05	1.375E-04	1.15	303.	13.2	22.8	48.1	93.4	47.9
813.	1.620E-05	1.328E-04	1.15	303.	12.5	23.3	48.9	93.5	46.6
838.	1.566E-05	1.284E-04	1.15	303.	11.9	23.8	49.7	93.6	45.2
864.	1.516E-05	1.242E-04	1.15	303.	11.2	24.3	50.5	93.6	43.8
877.	1.492E-05	1.222E-04	1.15	303.	10.9	24.6	50.8	93.7	43.0
902.	1.446E-05	1.185E-04	1.15	303.	10.2	25.1	51.6	93.7	41.5
915.	1.424E-05	1.167E-04	1.15	303.	9.87	25.3	51.9	93.8	40.8
941.	1.382E-05	1.132E-04	1.15	303.	9.22	25.8	52.7	93.8	39.2
954.	1.362E-05	1.116E-04	1.15	303.	8.90	26.1	53.0	93.8	38.4
979.	1.323E-05	1.084E-04	1.15	303.	8.27	26.6	53.7	93.8	36.8
992.	1.305E-05	1.069E-04	1.15	303.	7.95	26.8	54.1	93.8	35.9
1.018E+03	1.269E-05	1.040E-04	1.15	303.	7.33	27.3	54.8	93.9	34.1
1.030E+03	1.252E-05	1.026E-04	1.15	303.	7.03	27.6	55.1	93.9	33.2
1.056E+03	1.219E-05	9.993E-05	1.15	303.	6.42	28.0	55.8	93.9	31.3
1.069E+03	1.204E-05	9.864E-05	1.15	303.	6.12	28.3	56.2	93.9	30.4
1.094E+03	1.173E-05	9.615E-05	1.15	303.	5.52	28.8	56.9	93.9	28.3

1.107E+03	1.159E-05	9.495E-05	1.15	303.	5.22	29.0	57.2	93.8	27.2
1.133E+03	1.130E-05	9.264E-05	1.15	303.	4.63	29.5	57.8	93.8	25.0
1.146E+03	1.117E-05	9.152E-05	1.15	303.	4.34	29.7	58.2	93.8	23.8
1.158E+03	1.104E-05	9.043E-05	1.15	303.	4.05	29.9	58.5	93.8	22.5
1.171E+03	1.091E-05	8.937E-05	1.15	303.	3.77	30.2	58.8	93.8	21.2
1.184E+03	1.078E-05	8.833E-05	1.15	303.	3.48	30.4	59.2	93.8	19.8
1.197E+03	1.066E-05	8.731E-05	1.15	303.	3.19	30.6	59.5	93.7	18.3
1.210E+03	1.053E-05	8.632E-05	1.15	303.	2.91	30.9	59.8	93.7	16.6
1.222E+03	1.042E-05	8.535E-05	1.15	303.	2.63	31.1	60.1	93.7	14.9
1.235E+03	1.030E-05	8.439E-05	1.15	303.	2.35	31.3	60.4	93.7	12.9
1.248E+03	1.019E-05	8.346E-05	1.15	303.	2.07	31.5	60.7	93.6	10.5
1.261E+03	1.007E-05	8.255E-05	1.15	303.	1.79	31.8	61.1	93.6	7.32
1.274E+03	9.966E-06	8.166E-05	1.15	303.	1.52	32.0	61.4	93.6	
1.286E+03	9.859E-06	8.079E-05	1.15	303.	1.24	32.2	61.7	93.6	
1.299E+03	9.755E-06	7.994E-05	1.15	303.	0.970	32.4	62.0	93.5	
1.312E+03	9.653E-06	7.910E-05	1.15	303.	0.698	32.7	62.3	93.5	
1.325E+03	9.553E-06	7.828E-05	1.15	303.	0.427	32.9	62.6	93.5	
1.338E+03	9.455E-06	7.748E-05	1.15	303.	0.158	33.1	62.9	93.4	
1.350E+03	9.359E-06	7.669E-05	1.15	303.	0.000E-01	33.3	63.1	93.3	
1.376E+03	9.131E-06	7.482E-05	1.15	303.	0.000E-01	33.8	63.7	93.7	
1.402E+03	8.823E-06	7.230E-05	1.15	303.	0.000E-01	34.2	64.3	93.8	
1.440E+03	8.452E-06	6.926E-05	1.15	303.	0.000E-01	34.9	65.2	94.2	
1.478E+03	8.104E-06	6.641E-05	1.15	303.	0.000E-01	35.5	66.1	94.5	
1.517E+03	7.778E-06	6.373E-05	1.15	303.	0.000E-01	36.2	67.0	94.8	
1.555E+03	7.471E-06	6.122E-05	1.15	303.	0.000E-01	36.8	67.9	95.1	
1.594E+03	7.182E-06	5.885E-05	1.15	303.	0.000E-01	37.4	68.8	95.4	
1.632E+03	6.910E-06	5.662E-05	1.15	303.	0.000E-01	38.1	69.7	95.7	
1.670E+03	6.654E-06	5.452E-05	1.15	303.	0.000E-01	38.7	70.6	95.9	
1.709E+03	6.411E-06	5.254E-05	1.15	303.	0.000E-01	39.3	71.5	96.2	
1.747E+03	6.182E-06	5.066E-05	1.15	303.	0.000E-01	39.9	72.4	96.4	
1.786E+03	5.965E-06	4.888E-05	1.15	303.	0.000E-01	40.6	73.3	96.6	
1.837E+03	5.737E-06	4.701E-05	1.15	303.	0.000E-01	41.4	74.5	97.0	
1.875E+03	5.543E-06	4.542E-05	1.15	303.	0.000E-01	42.0	75.3	97.2	
1.914E+03	5.359E-06	4.392E-05	1.15	303.	0.000E-01	42.6	76.2	97.3	
1.952E+03	5.185E-06	4.248E-05	1.15	303.	0.000E-01	43.2	77.1	97.5	
2.003E+03	4.965E-06	4.068E-05	1.15	303.	0.000E-01	44.0	78.3	97.6	
2.054E+03	4.759E-06	3.900E-05	1.15	303.	0.000E-01	44.8	79.5	97.7	
2.106E+03	4.566E-06	3.741E-05	1.15	303.	0.000E-01	45.6	80.7	97.8	
2.157E+03	4.384E-06	3.593E-05	1.15	303.	0.000E-01	46.3	81.8	97.9	
2.208E+03	4.213E-06	3.453E-05	1.15	303.	0.000E-01	47.1	83.0	97.9	
2.259E+03	4.053E-06	3.321E-05	1.15	303.	0.000E-01	47.9	84.2	97.8	
2.310E+03	3.901E-06	3.197E-05	1.15	303.	0.000E-01	48.6	85.3	97.8	
2.362E+03	3.758E-06	3.079E-05	1.15	303.	0.000E-01	49.4	86.5	97.7	
2.413E+03	3.623E-06	2.969E-05	1.15	303.	0.000E-01	50.1	87.7	97.6	
2.464E+03	3.495E-06	2.864E-05	1.15	303.	0.000E-01	50.9	88.8	97.5	

2.515E+03	3.374E-06	2.764E-05	1.15	303.	0.000E-01	51.6	90.0	97.3
2.579E+03	3.231E-06	2.647E-05	1.15	303.	0.000E-01	52.5	91.5	97.0
2.643E+03	3.097E-06	2.538E-05	1.15	303.	0.000E-01	53.5	92.9	96.7
2.707E+03	2.972E-06	2.435E-05	1.15	303.	0.000E-01	54.4	94.3	96.3
2.771E+03	2.854E-06	2.339E-05	1.15	303.	0.000E-01	55.3	95.8	95.8
2.835E+03	2.743E-06	2.248E-05	1.15	303.	0.000E-01	56.2	97.2	95.3
2.899E+03	2.638E-06	2.162E-05	1.15	303.	0.000E-01	57.1	98.7	94.8
2.963E+03	2.540E-06	2.081E-05	1.15	303.	0.000E-01	57.9	100.	94.1
3.027E+03	2.447E-06	2.005E-05	1.15	303.	0.000E-01	58.8	102.	93.4
3.104E+03	2.342E-06	1.919E-05	1.15	303.	0.000E-01	59.9	103.	92.5
3.181E+03	2.244E-06	1.839E-05	1.15	303.	0.000E-01	60.9	105.	91.5
3.258E+03	2.152E-06	1.763E-05	1.15	303.	0.000E-01	61.9	107.	90.4
3.334E+03	2.065E-06	1.692E-05	1.15	303.	0.000E-01	62.9	108.	89.2
3.411E+03	1.984E-06	1.626E-05	1.15	303.	0.000E-01	63.9	110.	87.9
3.488E+03	1.907E-06	1.563E-05	1.15	303.	0.000E-01	65.0	112.	86.4
3.565E+03	1.835E-06	1.504E-05	1.15	303.	0.000E-01	66.0	114.	84.9
3.642E+03	1.767E-06	1.448E-05	1.15	303.	0.000E-01	66.9	115.	83.2
3.718E+03	1.703E-06	1.395E-05	1.15	303.	0.000E-01	67.9	117.	81.4
3.795E+03	1.642E-06	1.346E-05	1.15	303.	0.000E-01	68.9	119.	79.4
3.872E+03	1.585E-06	1.299E-05	1.15	303.	0.000E-01	69.9	120.	77.2
3.949E+03	1.530E-06	1.254E-05	1.15	303.	0.000E-01	70.8	122.	74.9
4.026E+03	1.478E-06	1.211E-05	1.15	303.	0.000E-01	71.8	124.	72.4
4.102E+03	1.429E-06	1.171E-05	1.15	303.	0.000E-01	72.8	125.	69.7
4.179E+03	1.383E-06	1.133E-05	1.15	303.	0.000E-01	73.7	127.	66.7
4.256E+03	1.338E-06	1.097E-05	1.15	303.	0.000E-01	74.6	129.	63.5
4.333E+03	1.296E-06	1.062E-05	1.15	303.	0.000E-01	75.6	130.	59.9
4.410E+03	1.256E-06	1.029E-05	1.15	303.	0.000E-01	76.5	132.	56.0
4.486E+03	1.217E-06	9.976E-06	1.15	303.	0.000E-01	77.4	134.	51.6
4.563E+03	1.181E-06	9.676E-06	1.15	303.	0.000E-01	78.4	135.	46.5
4.640E+03	1.146E-06	9.389E-06	1.15	303.	0.000E-01	79.3	137.	40.7
4.717E+03	1.112E-06	9.116E-06	1.15	303.	0.000E-01	80.2	139.	33.6
4.794E+03	1.080E-06	8.854E-06	1.15	303.	0.000E-01	81.1	140.	24.1
4.870E+03	1.050E-06	8.603E-06	1.15	303.	0.000E-01	82.0	142.	3.67
4.947E+03	1.021E-06	8.363E-06	1.15	303.	0.000E-01	82.9	144.	
5.024E+03	9.926E-07	8.133E-06	1.15	303.	0.000E-01	83.8	145.	
5.101E+03	9.657E-07	7.913E-06	1.15	303.	0.000E-01	84.7	147.	
5.178E+03	9.399E-07	7.702E-06	1.15	303.	0.000E-01	85.5	149.	
5.254E+03	9.151E-07	7.499E-06	1.15	303.	0.000E-01	86.4	150.	
5.331E+03	8.913E-07	7.304E-06	1.15	303.	0.000E-01	87.3	152.	
5.408E+03	8.685E-07	7.116E-06	1.15	303.	0.000E-01	88.2	154.	
5.485E+03	8.465E-07	6.936E-06	1.15	303.	0.000E-01	89.0	155.	
5.562E+03	8.253E-07	6.763E-06	1.15	303.	0.000E-01	89.9	157.	
5.638E+03	8.050E-07	6.596E-06	1.15	303.	0.000E-01	90.7	158.	
5.715E+03	7.854E-07	6.436E-06	1.15	303.	0.000E-01	91.6	160.	
5.792E+03	7.665E-07	6.281E-06	1.15	303.	0.000E-01	92.4	162.	

```
5.869E+03  7.483E-07  6.132E-06  1.15    303.    0.000E-01  93.3    163.
5.946E+03  7.308E-07  5.988E-06  1.15    303.    0.000E-01  94.1    165.

6.022E+03  7.139E-07  5.850E-06  1.15    303.    0.000E-01  95.0    167.
6.099E+03  6.975E-07  5.716E-06  1.15    303.    0.000E-01  95.8    168.
6.176E+03  6.817E-07  5.586E-06  1.15    303.    0.000E-01  96.6    170.

6.253E+03  6.665E-07  5.462E-06  1.15    303.    0.000E-01  97.5    172.
6.330E+03  6.518E-07  5.341E-06  1.15    303.    0.000E-01  98.3    173.
6.406E+03  6.375E-07  5.224E-06  1.15    303.    0.000E-01  99.1    175.

6.483E+03  6.238E-07  5.111E-06  1.15    303.    0.000E-01  99.9    176.
6.560E+03  6.105E-07  5.002E-06  1.15    303.    0.000E-01  101.    178.
6.637E+03  5.976E-07  4.897E-06  1.15    303.    0.000E-01  102.    180.

6.714E+03  5.851E-07  4.794E-06  1.15    303.    0.000E-01  102.    181.
6.790E+03  5.730E-07  4.695E-06  1.15    303.    0.000E-01  103.    183.
6.867E+03  5.613E-07  4.599E-06  1.15    303.    0.000E-01  104.    184.

6.944E+03  5.499E-07  4.506E-06  1.15    303.    0.000E-01  105.    186.
7.021E+03  5.389E-07  4.416E-06  1.15    303.    0.000E-01  106.    188.
7.098E+03  5.283E-07  4.329E-06  1.15    303.    0.000E-01  106.    189.

7.136E+03  5.231E-07  4.286E-06  1.15    303.    0.000E-01  107.    190.
```

For the UFL of   1.00000E-03 mole percent, and the LFL of   1.05000E-04 mole percent:

The mass of contaminant between the UFL and LFL is: -114.06   kg.
The mass of contaminant above the LFL is: 577.44   kg.

6. Scenario 4: SLAB

Stability Class = F
Wind Speed $u = 2$ m/s
Roughness Length $z_0 = 0.3$ m
Averaging time $t_a = 1$ min
Source: Flashed Vapor—No Aerosol

problem input

```
wms   =       .071
cps   =    480.000
ts    =    239.000
qs    =       .246
as    =      1.000
avt   =     60.000
xffm  =   3000.000
zp(2) =      1.000
zp(3) =       .000
zp(4) =       .000
```

release gas properties

molecular weight of source gas (kg)	- wms =	.7091E-01
heat capacity at const. p. (j/kg-k)	- cps =	.4800E+03
temperature of source gas (k)	- ts =	.2390E+03
density of source gas (kg/m3)	- rhos =	.3615E+01

spill characteristics

mass source rate (kg/s)	- qs =	.2457E+00
source area (m2)	- as =	.1000E+01
vapor injection velocity (evaporation rate) (m/s)	- ws =	.6796E-01
source half width -.5*((qqs/ws)**.5) (m)	- bs =	.5000E+00

field parameters

concentration averaging time (s)	- avt =	.6000E+02
maximum downwind distrace (m)	- xffm =	.3000E+04
concentration measurement height (m)	- zp(1)=	.0000E+00
	- zp(2)=	.1000E+01
	- zp(3)=	.0000E+00
	- zp(4)=	.0000E+00

ambient meteorological properties

molecular weight of air (kg)	- wma =	.2896E-01
heat capacity of air at const. p. (j/kg-k)	- cpa =	.1006E+04
density of ambient air (kg/m3)	- rhoa =	.1165E+01
ambient measurement height (m)	- za =	.1000E+02
ambient atmospheric pressure (atm)	- pa =	.1000E+01
ambient wind speed (m/s)	- ua =	.2000E+01
ambient temperature (k)	- ta =	.3030E+03
ambient friction velocity (m/s)	- uastr =	.2191E+00
inverse monin-obukhov length (1/m)	- ala =	.4733E-01
surface roughness height (m)	- z0 =	.3000E+00

additional parameters

acceleration of gravity (m/s2)	- grav =	.9807E+01

```
gas constant (m3 -atm/mol- k)                    - rr   =   .8206E-04
von karman constant                              - xk   =   .4100E+00



source region

              [INTERMEDIATE OUTPUT REMOVED]


volume concentration of (x,z)

       x        z=  .00    z= 1.00    z=  .00    z=  .00
    -.5422E+00  .0000E+00  .0000E+00  .0000E+00  .0000E+00
    -.4337E+00  .3832E-01  .9198E-24  .0000E+00  .0000E+00
    -.3253E+00  .6393E-01  .3132E-15  .0000E+00  .0000E+00
    -.2169E+00  .8730E-01  .1302E-11  .0000E+00  .0000E+00
    -.1084E+00  .1093E+00  .1986E-09  .0000E+00  .0000E+00
     .8941E-07  .1300E+00  .6935E-08  .0000E+00  .0000E+00
     .1084E+00  .1495E+00  .1017E-06  .0000E+00  .0000E+00
     .2169E+00  .1680E+00  .8231E-06  .0000E+00  .0000E+00
     .3253E+00  .1856E+00  .4405E-05  .0000E+00  .0000E+00
     .4337E+00  .2024E+00  .1882E-04  .0000E+00  .0000E+00
     .5422E+00  .2182E+00  .1670E-03  .0000E+00  .0000E+00



    far field


        input source parameters      effective source parameters
           bs =  .5000E+00              bse =  .5422E+00
           ws =  .6796E-01              wse =  .5779E-01



              [INTERMEDIATE OUTPUT REMOVED]


volume concentration of (x,z)

       x        z=  .00    z= 1.00    z=  .00    z=  .00
    -.5422E+00  .0000E+00  .0000E+00  .0000E+00  .0000E+00
    -.4337E+00  .3832E-01  .9198E-24  .0000E+00  .0000E+00
    -.3253E+00  .6393E-01  .3132E-15  .0000E+00  .0000E+00
    -.2169E+00  .8730E-01  .1302E-11  .0000E+00  .0000E+00
    -.1084E+00  .1093E+00  .1986E-09  .0000E+00  .0000E+00
     .8941E-07  .1300E+00  .6935E-08  .0000E+00  .0000E+00
     .1084E+00  .1495E+00  .1017E-06  .0000E+00  .0000E+00
     .2169E+00  .1680E+00  .8231E-06  .0000E+00  .0000E+00
     .3253E+00  .1856E+00  .4405E-05  .0000E+00  .0000E+00
     .4337E+00  .2024E+00  .1882E-04  .0000E+00  .0000E+00
     .5422E+00  .2182E+00  .1670E-03  .0000E+00  .0000E+00
     .6029E+00  .2162E+00  .1485E-04  .0000E+00  .0000E+00
     .6758E+00  .2133E+00  .2131E-05  .0000E+00  .0000E+00
     .7635E+00  .2093E+00  .2313E-06  .0000E+00  .0000E+00
     .8689E+00  .2037E+00  .1696E-07  .0000E+00  .0000E+00
     .9957E+00  .1961E+00  .8410E-09  .0000E+00  .0000E+00
     .1148E+01  .1858E+00  .3290E-10  .0000E+00  .0000E+00
     .1331E+01  .1726E+00  .1472E-11  .0000E+00  .0000E+00
     .1552E+01  .1564E+00  .1383E-12  .0000E+00  .0000E+00
```

.1817E+01	.1376E+00	.5589E-13	.0000E+00	.0000E+00
.2135E+01	.1173E+00	.1621E-12	.0000E+00	.0000E+00
.2518E+01	.9675E-01	.3141E-11	.0000E+00	.0000E+00
.2978E+01	.7770E-01	.1939E-09	.0000E+00	.0000E+00
.3531E+01	.6128E-01	.1387E-07	.0000E+00	.0000E+00
.4197E+01	.4795E-01	.5496E-06	.0000E+00	.0000E+00
.4997E+01	.3752E-01	.9239E-05	.0000E+00	.0000E+00
.5958E+01	.2951E-01	.6965E-04	.0000E+00	.0000E+00
.7114E+01	.2334E-01	.2795E-03	.0000E+00	.0000E+00
.8504E+01	.1858E-01	.7091E-03	.0000E+00	.0000E+00
.1018E+02	.1485E-01	.1302E-02	.0000E+00	.0000E+00
.1218E+02	.1190E-01	.1908E-02	.0000E+00	.0000E+00
.1460E+02	.9545E-02	.2383E-02	.0000E+00	.0000E+00
.1750E+02	.7650E-02	.2654E-02	.0000E+00	.0000E+00
.2100E+02	.6118E-02	.2720E-02	.0000E+00	.0000E+00
.2519E+02	.4877E-02	.2618E-02	.0000E+00	.0000E+00
.3024E+02	.3871E-02	.2403E-02	.0000E+00	.0000E+00
.3631E+02	.3058E-02	.2123E-02	.0000E+00	.0000E+00
.4360E+02	.2403E-02	.1820E-02	.0000E+00	.0000E+00
.5237E+02	.1879E-02	.1522E-02	.0000E+00	.0000E+00
.6292E+02	.1460E-02	.1246E-02	.0000E+00	.0000E+00
.7559E+02	.1127E-02	.1001E-02	.0000E+00	.0000E+00
.9083E+02	.8651E-03	.7924E-03	.0000E+00	.0000E+00
.1092E+03	.6594E-03	.6183E-03	.0000E+00	.0000E+00
.1312E+03	.4991E-03	.4763E-03	.0000E+00	.0000E+00
.1577E+03	.3751E-03	.3626E-03	.0000E+00	.0000E+00
.1895E+03	.2800E-03	.2733E-03	.0000E+00	.0000E+00
.2278E+03	.2076E-03	.2041E-03	.0000E+00	.0000E+00
.2738E+03	.1532E-03	.1514E-03	.0000E+00	.0000E+00
.3292E+03	.1127E-03	.1117E-03	.0000E+00	.0000E+00
.3957E+03	.8278E-04	.8228E-04	.0000E+00	.0000E+00
.4757E+03	.6087E-04	.6061E-04	.0000E+00	.0000E+00
.5718E+03	.4489E-04	.4476E-04	.0000E+00	.0000E+00
.6874E+03	.3326E-04	.3319E-04	.0000E+00	.0000E+00
.8264E+03	.2479E-04	.2475E-04	.0000E+00	.0000E+00
.9936E+03	.1859E-04	.1857E-04	.0000E+00	.0000E+00
.1194E+04	.1404E-04	.1402E-04	.0000E+00	.0000E+00
.1436E+04	.1067E-04	.1066E-04	.0000E+00	.0000E+00
.1726E+04	.8171E-05	.8166E-05	.0000E+00	.0000E+00
.2076E+04	.6299E-05	.6296E-05	.0000E+00	.0000E+00
.2495E+04	.4889E-05	.4887E-05	.0000E+00	.0000E+00
.3000E+04	.3821E-05	.3819E-05	.0000E+00	.0000E+00

7. Scenario 4: CAMEO

Stability Class = F
Wind Speed $u = 2$ m/s
Roughness = "Homogeneous Forest, Suburb Full
Obstacle" (≈ 0.3 m)
Source: Total Mass Released as a Gas

AIR MODEL 3.3

CHEMICAL NAME : CHLORINE

TLV-TWA=   1.00 PPM   IDLH=   25.00 PPM
V. P.= 387.40 MM (HG) at -55.0 F   MOL. WT.=   70.91
BOILING POINT TEMPERATURE IS   -30.3 F
SATURATION CONC. AT 1 ATM. AND   30.0 C IS 1000000 PPM

WIND SPEED= 4 KNOTS FROM 90.0 DEG. TRUE
AMBIENT TEMPERATURE IS 30.0 DEG. C
NO INVERSION PRESENT   /   STAB. CLASS=F
GROUND ROUGHNESS IS HOMOGENEOUS FOREST, SUBURB FULL OBSTACLE

USER INPUT SOURCE STRENGTH DIRECTLY
SOURCE STRENGTH IS :   1000.0 GM/SEC

FOR A CONTINUOUS SOURCE:
DOWNWIND IDLH DIST & TRAVEL TIME :   1.5 MILE 19.3 MIN.
DOWNWIND TLV-TWA DIST & TRAVEL TIME :   13.058 MILES & 170.2 MIN.

DILUTION CONTOURS

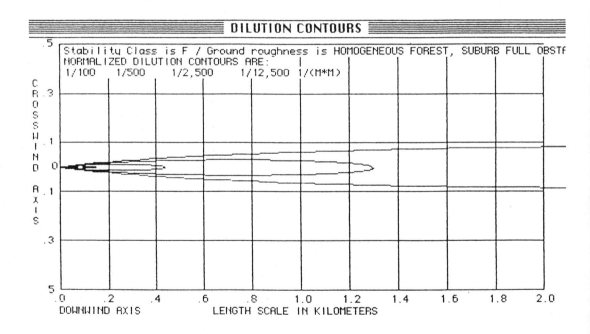

Stability Class is F / Ground roughness is HOMOGENEOUS FOREST, SUBURB FULL OBSTF
NORMALIZED DILUTION CONTOURS ARE:
  1/100   1/500   1/2,500   1/12,500 1/(M*M)

CROSSWIND AXIS

DOWNWIND AXIS   LENGTH SCALE IN KILOMETERS

8. Scenario 5: DEGADIS

Stability Class = F
Wind Speed $u = 2$ m/s
Roughness Length $z_0 = 0.3$ m
Averaging time $t_a = 1$ min

0   TITLE BLOCK

SC=F WS=2 Z0=.3 TIME=1 MIN


```
         Wind velocity at reference height              2.00  m/s
         Reference height                              10.00  m
0        Surface roughness length             0.300          m
0        Pasquill Stability class                F
0        Monin-Obukhov length                    21.1       m
         Gaussian distribution constants  Delta    0.04040   m
                                          Beta      0.90000
0        Wind velocity power law constant Alpha     0.53216
         Friction velocity                          0.12155  m/s
0        Ambient Temperature                      303.15  K
         Ambient Pressure                           1.000  atm
         Ambient Absolute Humidity           1.389E-02  kg/kg BDA
         Ambient Relative Humidity              50.00  %
```

Adiabatic Mixing:	Mole fraction	CONCENTRATION OF C kg/m**3	GAS DENSITY kg/m**3	Enthalpy J/kg	Temperature K
	0.00000	0.00000	1.15593	0.00000E-01	303.15
	0.06598	0.15423	1.23386	-10.750	303.14
	0.14151	0.33076	1.32306	-21.501	303.13
	0.22881	0.53481	1.42615	-32.251	303.11
	0.33088	0.77333	1.54666	-43.001	303.10
	0.45180	1.05588	1.68942	-53.751	303.08
	0.59734	1.39591	1.86121	-64.502	303.05
	0.77586	1.81289	2.07188	-75.252	303.03
	1.00000	2.33630	2.33630	-86.002	303.00

0   Specified Gas Properties:

```
         Molecular weight:                              58.080
         Storage temperature:                          303.00    K
         Density at storage temperature and ambient pressure:   2.3363   kg/m**3
         Mean heat capacity constant:                  0.00000E-01
```

```
        Mean heat capacity power:                         1.0000
        Upper mole fraction contour:                      0.15000
        Lower mole fraction contour:                      2.00000E-05
        Height for isopleths:                             0.00000E-01m
```

Source input data points

```
              Initial mass in cloud:    0.00000E-01

                         TIME        SOURCE STRENGTH      SOURCE RADIUS
                          s               kg/s                m
                      0.00000E-01        0.84190            10.290
                        6023.0           0.84190            10.290
                        6024.0         0.00000E-01        0.00000E-01
                        6025.0         0.00000E-01        0.00000E-01
```

```
0    Calculation procedure for ALPHA:  1
0    Entrainment prescription for PHI:  3
0    Layer thickness ratio used for average depth:   2.1500
0    Air entrainment coefficient used: 0.590
0    Gravity slumping velocity coefficient used: 1.150
0    NON Isothermal calculation
0    Heat transfer not included
0    Water transfer not included
```

```
                    *****              CALCULATED SOURCE PARAMETERS            *****

    Time     Gas Radius    Height      Qstar      SZ(x=L/2.)  Mole frac C    Density    Temperature   Rich No.
    sec          m           m       kg/m**2/s       m                      kg/m**3        K

   60.2300    10.2900    0.000000E-01  2.530928E-03   1.48048   1.787365E-02   1.17704      303.147    0.000000E-01
   301.150    10.2900    0.000000E-01  2.530928E-03   1.48048   1.787365E-02   1.17704      303.147    0.000000E-01
0Source strength [kg/s] :              0.84190       Equivalent Primary source radius [m] :     10.290
 Equivalent Primary source length [m] :   18.239    Equivalent Primary source width [m] :      18.239

Secondary source concentration [kg/m**3] :   4.17810E-02  Secondary source SZ [m] :            1.4805

Contaminant flux rate:    2.53093E-03
```

```
Secondary source mass fractions... contaminant: 3.549655E-02   air:  0.95129
          Enthalpy:   -3.0528        Density:   1.1770
Secondary source length [m] :           18.239     Secondary source half-width [m] :        9.1193
```

```
0 Distance    Mole   Concentration Density Temperature   Half      Sz        Sy     Width at z= 0.00 m to:
             Fraction                                    Width                      2.000E-03mole% 15.0    mole%
   (m)                (kg/m**3)  (kg/m**3)    (K)         (m)       (m)       (m)        (m)         (m)

   9.12     1.787E-02  4.178E-02   1.18       303.        9.12      1.48      5.97       24.7
   10.7     1.692E-02  3.956E-02   1.18       303.        9.01      1.53      6.09       24.8
   12.3     1.606E-02  3.754E-02   1.17       303.        8.91      1.59      6.20       25.0

   15.5     1.454E-02  3.399E-02   1.17       303.        8.71      1.69      6.43       25.2
   18.7     1.326E-02  3.100E-02   1.17       303.        8.52      1.80      6.65       25.5
   21.9     1.217E-02  2.844E-02   1.17       303.        8.33      1.90      6.86       25.7

   28.3     1.040E-02  2.432E-02   1.17       303.        7.97      2.11      7.26       26.1
   34.7     9.042E-03  2.114E-02   1.17       303.        7.63      2.31      7.65       26.5
```

47.5	7.106E-03	1.661E-02	1.16	303.	7.00	2.70	8.36	27.3
60.3	5.802E-03	1.356E-02	1.16	303.	6.42	3.09	9.02	27.9
73.1	4.871E-03	1.139E-02	1.16	303.	5.87	3.46	9.63	28.4
85.9	4.179E-03	9.769E-03	1.16	303.	5.36	3.82	10.2	29.0
98.7	3.646E-03	8.523E-03	1.16	303.	4.88	4.18	10.8	29.4
112.	3.224E-03	7.538E-03	1.16	303.	4.42	4.53	11.3	29.8
124.	2.884E-03	6.741E-03	1.16	303.	3.98	4.87	11.8	30.2
137.	2.603E-03	6.085E-03	1.16	303.	3.56	5.21	12.2	30.6
150.	2.368E-03	5.536E-03	1.16	303.	3.15	5.54	12.7	30.9
163.	2.170E-03	5.072E-03	1.16	303.	2.76	5.86	13.1	31.2
176.	1.999E-03	4.674E-03	1.16	303.	2.38	6.18	13.6	31.5
188.	1.852E-03	4.329E-03	1.16	303.	2.02	6.50	14.0	31.8
201.	1.723E-03	4.029E-03	1.16	303.	1.66	6.81	14.4	32.0
214.	1.610E-03	3.764E-03	1.16	303.	1.31	7.12	14.8	32.3
227.	1.510E-03	3.530E-03	1.16	303.	0.974	7.43	15.2	32.5
240.	1.421E-03	3.321E-03	1.16	303.	0.644	7.73	15.5	32.7
252.	1.341E-03	3.134E-03	1.16	303.	0.322	8.03	15.9	32.9
265.	1.268E-03	2.965E-03	1.16	303.	6.877E-03	8.32	16.2	33.1
278.	1.203E-03	2.812E-03	1.16	303.	0.000E-01	8.62	16.3	32.9
284.	1.168E-03	2.729E-03	1.16	303.	0.000E-01	8.76	16.4	33.1
291.	1.119E-03	2.617E-03	1.16	303.	0.000E-01	8.91	16.6	33.3
304.	1.044E-03	2.440E-03	1.16	303.	0.000E-01	9.20	17.0	33.7
316.	9.751E-04	2.279E-03	1.16	303.	0.000E-01	9.49	17.3	34.1
329.	9.127E-04	2.134E-03	1.16	303.	0.000E-01	9.78	17.7	34.5
342.	8.559E-04	2.001E-03	1.16	303.	0.000E-01	10.1	18.0	34.9
355.	8.041E-04	1.880E-03	1.16	303.	0.000E-01	10.4	18.4	35.3
368.	7.566E-04	1.769E-03	1.16	303.	0.000E-01	10.6	18.7	35.6
380.	7.130E-04	1.667E-03	1.16	303.	0.000E-01	10.9	19.0	36.0
393.	6.730E-04	1.573E-03	1.16	303.	0.000E-01	11.2	19.4	36.4
406.	6.362E-04	1.487E-03	1.16	303.	0.000E-01	11.5	19.7	36.7
419.	6.022E-04	1.408E-03	1.16	303.	0.000E-01	11.8	20.1	37.0
432.	5.707E-04	1.334E-03	1.16	303.	0.000E-01	12.1	20.4	37.4
444.	5.416E-04	1.266E-03	1.16	303.	0.000E-01	12.4	20.8	37.7
457.	5.146E-04	1.203E-03	1.16	303.	0.000E-01	12.7	21.1	38.0
470.	4.895E-04	1.144E-03	1.16	303.	0.000E-01	12.9	21.4	38.3
483.	4.661E-04	1.090E-03	1.16	303.	0.000E-01	13.2	21.8	38.7
496.	4.443E-04	1.039E-03	1.16	303.	0.000E-01	13.5	22.1	39.0
508.	4.240E-04	9.911E-04	1.16	303.	0.000E-01	13.8	22.5	39.3
521.	4.050E-04	9.467E-04	1.16	303.	0.000E-01	14.1	22.8	39.5
534.	3.872E-04	9.051E-04	1.16	303.	0.000E-01	14.3	23.1	39.8
547.	3.705E-04	8.661E-04	1.16	303.	0.000E-01	14.6	23.5	40.1
572.	3.402E-04	7.952E-04	1.16	303.	0.000E-01	15.2	24.1	40.7
585.	3.264E-04	7.629E-04	1.16	303.	0.000E-01	15.5	24.5	40.9
611.	3.011E-04	7.038E-04	1.16	303.	0.000E-01	16.0	25.2	41.4
624.	2.895E-04	6.767E-04	1.16	303.	0.000E-01	16.3	25.5	41.7

649.	2.682E-04	6.270E-04	1.16	303.	0.000E-01	16.8	26.2	42.1
662.	2.584E-04	6.041E-04	1.16	303.	0.000E-01	17.1	26.5	42.4
688.	2.404E-04	5.619E-04	1.16	303.	0.000E-01	17.6	27.2	42.8
700.	2.320E-04	5.424E-04	1.16	303.	0.000E-01	17.9	27.5	43.0
726.	2.166E-04	5.063E-04	1.16	303.	0.000E-01	18.4	28.1	43.4
739.	2.094E-04	4.895E-04	1.16	303.	0.000E-01	18.7	28.5	43.6
764.	1.961E-04	4.584E-04	1.16	303.	0.000E-01	19.2	29.1	44.0
777.	1.899E-04	4.439E-04	1.16	303.	0.000E-01	19.5	29.5	44.2
803.	1.784E-04	4.169E-04	1.16	303.	0.000E-01	20.0	30.1	44.6
816.	1.730E-04	4.043E-04	1.16	303.	0.000E-01	20.3	30.5	44.7
841.	1.629E-04	3.808E-04	1.16	303.	0.000E-01	20.8	31.1	45.0
854.	1.582E-04	3.698E-04	1.16	303.	0.000E-01	21.1	31.4	45.2
880.	1.493E-04	3.491E-04	1.16	303.	0.000E-01	21.6	32.1	45.5
892.	1.452E-04	3.394E-04	1.16	303.	0.000E-01	21.8	32.4	45.6
918.	1.374E-04	3.211E-04	1.16	303.	0.000E-01	22.3	33.1	45.9
931.	1.337E-04	3.126E-04	1.16	303.	0.000E-01	22.6	33.4	46.0
956.	1.268E-04	2.964E-04	1.16	303.	0.000E-01	23.1	34.0	46.2
969.	1.235E-04	2.888E-04	1.16	303.	0.000E-01	23.4	34.4	46.4
995.	1.174E-04	2.744E-04	1.16	303.	0.000E-01	23.9	35.0	46.6
1.008E+03	1.145E-04	2.676E-04	1.16	303.	0.000E-01	24.1	35.3	46.7
1.033E+03	1.090E-04	2.547E-04	1.16	303.	0.000E-01	24.6	36.0	46.8
1.046E+03	1.064E-04	2.486E-04	1.16	303.	0.000E-01	24.8	36.3	46.9
1.072E+03	1.014E-04	2.371E-04	1.16	303.	0.000E-01	25.3	36.9	47.1
1.084E+03	9.908E-05	2.316E-04	1.16	303.	0.000E-01	25.6	37.3	47.1
1.110E+03	9.463E-05	2.212E-04	1.16	303.	0.000E-01	26.1	37.9	47.2
1.123E+03	9.251E-05	2.163E-04	1.16	303.	0.000E-01	26.3	38.2	47.3
1.148E+03	8.849E-05	2.069E-04	1.16	303.	0.000E-01	26.8	38.9	47.4
1.174E+03	8.472E-05	1.981E-04	1.16	303.	0.000E-01	27.3	39.5	47.4
1.200E+03	8.119E-05	1.898E-04	1.16	303.	0.000E-01	27.8	40.1	47.5
1.225E+03	7.788E-05	1.820E-04	1.16	303.	0.000E-01	28.2	40.8	47.5
1.251E+03	7.476E-05	1.748E-04	1.16	303.	0.000E-01	28.7	41.4	47.5
1.276E+03	7.182E-05	1.679E-04	1.16	303.	0.000E-01	29.2	42.0	47.5
1.302E+03	6.906E-05	1.614E-04	1.16	303.	0.000E-01	29.6	42.7	47.5
1.328E+03	6.645E-05	1.553E-04	1.16	303.	0.000E-01	30.1	43.3	47.4
1.353E+03	6.398E-05	1.496E-04	1.16	303.	0.000E-01	30.6	43.9	47.4
1.379E+03	6.165E-05	1.441E-04	1.16	303.	0.000E-01	31.0	44.5	47.3
1.404E+03	5.945E-05	1.390E-04	1.16	303.	0.000E-01	31.5	45.2	47.1
1.430E+03	5.736E-05	1.341E-04	1.16	303.	0.000E-01	31.9	45.8	47.0
1.456E+03	5.538E-05	1.295E-04	1.16	303.	0.000E-01	32.4	46.4	46.9
1.481E+03	5.350E-05	1.251E-04	1.16	303.	0.000E-01	32.8	47.1	46.7
1.507E+03	5.171E-05	1.209E-04	1.16	303.	0.000E-01	33.3	47.7	46.5
1.532E+03	5.001E-05	1.169E-04	1.16	303.	0.000E-01	33.7	48.3	46.2
1.558E+03	4.840E-05	1.131E-04	1.16	303.	0.000E-01	34.2	48.9	46.0
1.584E+03	4.686E-05	1.095E-04	1.16	303.	0.000E-01	34.6	49.5	45.7
1.609E+03	4.539E-05	1.061E-04	1.16	303.	0.000E-01	35.1	50.2	45.4

1.635E+03	4.400E-05	1.028E-04	1.16	303.	0.000E-01	35.5	50.8	45.1
1.660E+03	4.266E-05	9.972E-05	1.16	303.	0.000E-01	35.9	51.4	44.7
1.686E+03	4.139E-05	9.675E-05	1.16	303.	0.000E-01	36.4	52.0	44.4
1.712E+03	4.017E-05	9.390E-05	1.16	303.	0.000E-01	36.8	52.6	44.0
1.750E+03	3.844E-05	8.986E-05	1.16	303.	0.000E-01	37.4	53.6	43.3
1.788E+03	3.682E-05	8.607E-05	1.16	303.	0.000E-01	38.1	54.5	42.6
1.827E+03	3.530E-05	8.252E-05	1.16	303.	0.000E-01	38.7	55.4	41.8
1.865E+03	3.388E-05	7.919E-05	1.16	303.	0.000E-01	39.4	56.3	40.9
1.904E+03	3.253E-05	7.605E-05	1.16	303.	0.000E-01	40.0	57.2	39.9
1.942E+03	3.127E-05	7.310E-05	1.16	303.	0.000E-01	40.6	58.2	38.9
1.993E+03	2.993E-05	6.996E-05	1.16	303.	0.000E-01	41.4	59.4	37.7
2.032E+03	2.881E-05	6.735E-05	1.16	303.	0.000E-01	42.0	60.3	36.4
2.070E+03	2.776E-05	6.489E-05	1.16	303.	0.000E-01	42.7	61.2	35.0
2.108E+03	2.676E-05	6.256E-05	1.16	303.	0.000E-01	43.3	62.1	33.5
2.147E+03	2.582E-05	6.035E-05	1.16	303.	0.000E-01	43.9	63.0	31.8
2.185E+03	2.492E-05	5.826E-05	1.16	303.	0.000E-01	44.5	63.9	30.0
2.224E+03	2.408E-05	5.628E-05	1.16	303.	0.000E-01	45.1	64.8	27.9
2.262E+03	2.327E-05	5.440E-05	1.16	303.	0.000E-01	45.7	65.7	25.6
2.300E+03	2.250E-05	5.261E-05	1.16	303.	0.000E-01	46.3	66.6	22.9
2.352E+03	2.154E-05	5.035E-05	1.16	303.	0.000E-01	47.0	67.8	18.5
2.403E+03	2.064E-05	4.825E-05	1.16	303.	0.000E-01	47.8	69.0	12.2
2.454E+03	1.979E-05	4.627E-05	1.16	303.	0.000E-01	48.6	70.2	
2.505E+03	1.900E-05	4.441E-05	1.16	303.	0.000E-01	49.4	71.4	
2.556E+03	1.825E-05	4.266E-05	1.16	303.	0.000E-01	50.1	72.6	
2.608E+03	1.755E-05	4.102E-05	1.16	303.	0.000E-01	50.9	73.8	
2.659E+03	1.688E-05	3.947E-05	1.16	303.	0.000E-01	51.6	75.0	
2.710E+03	1.626E-05	3.800E-05	1.16	303.	0.000E-01	52.4	76.2	
2.761E+03	1.566E-05	3.662E-05	1.16	303.	0.000E-01	53.1	77.3	
2.812E+03	1.510E-05	3.531E-05	1.16	303.	0.000E-01	53.9	78.5	
2.864E+03	1.457E-05	3.407E-05	1.16	303.	0.000E-01	54.6	79.7	
2.928E+03	1.395E-05	3.261E-05	1.16	303.	0.000E-01	55.5	81.2	
2.992E+03	1.337E-05	3.124E-05	1.16	303.	0.000E-01	56.4	82.6	
3.056E+03	1.282E-05	2.996E-05	1.16	303.	0.000E-01	57.3	84.1	
3.120E+03	1.230E-05	2.876E-05	1.16	303.	0.000E-01	58.2	85.6	
3.184E+03	1.182E-05	2.763E-05	1.16	303.	0.000E-01	59.1	87.0	
3.248E+03	1.136E-05	2.656E-05	1.16	303.	0.000E-01	60.0	88.5	
3.312E+03	1.093E-05	2.556E-05	1.16	303.	0.000E-01	60.8	89.9	
3.376E+03	1.053E-05	2.461E-05	1.16	303.	0.000E-01	61.7	91.4	
3.440E+03	1.014E-05	2.371E-05	1.16	303.	0.000E-01	62.6	92.8	
3.504E+03	9.782E-06	2.287E-05	1.16	303.	0.000E-01	63.4	94.3	

For the UFL of   15.000   mole percent, and the LFL of   2.00000E-03 mole percent:

The mass of contaminant between the UFL and LFL is: -262.14   kg.
The mass of contaminant above the LFL is: 238.01   kg.

116

9. Scenario 5: SPILLS

Stability Class = F
Wind Speed u = 2 m/s

```
***   F2
** The concentrations contained in this file are the
** maximum concentrations obtained at the centerline of
** the plume at the distances requested and at the
** elevation and elasped time (if unsteady-state) as
** specified by the user.
** This model is not appropriate for calculating
** concentrations at receptors less than 100 meters
** from the source.
** STABILITY CLASS IS         6
** MOLECULAR WEIGHT IS    0.0000
** WIND SPEED IS (M/SEC)    2.000
** EMISSION RATE IS (M/SEC)   0.0000
** AT TIME' (MIN)   0.0000
** EMISSION RATE IS (M/SEC)   0.8419
** AT TIME' (MIN)   0.1000E-04
** EMISSION RATE IS (M/SEC)   0.8419
** AT TIME' (MIN)   390.9
** EMISSION RATE IS (M/SEC)   0.0000
** AT TIME' (MIN)   390.9
** AMBIENT TEMPERTURE IS (DEGREES C)    30.00
** STACK HEIGHT IS (M)   0.0000
** AREA IS (M)    332.7
** DISTANCE   CONCENTRATION
** METERS       MG/M**3
** DISTANCE   CONCENTRATION
**   FEET        PPM
```

DISTANCE (FEET)	CONCENTRATION (PPM)	DISTANCE (FEET)	CONCENTRATION (PPM)
100.0	7353.		
200.0	2888.		
300.0	1616.		
400.0	1052.		
500.0	748.0		
600.0	563.4		
700.0	442.0		
800.0	362.4		
900.0	303.7		
1000.0	259.1		
1100.0	225.4		
1200.0	198.4		
1300.0	176.3		
1400.0	158.0		
1500.0	142.7		
1600.0	129.6		
1700.0	118.5		
1800.0	108.8		
1900.0	100.4		
2000.0	92.95		
2100.0	86.78		
2200.0	81.27		
2300.0	76.33		
2400.0	71.88		
2500.0	67.85		
2600.0	64.18	3400.0	44.32
2700.0	60.84	3500.0	42.63
2800.0	57.79	3600.0	41.05
2900.0	54.99	3700.0	39.56
3000.0	52.40	3800.0	38.17
3100.0	50.15	3900.0	36.87
3200.0	48.07	4000.0	35.63
3300.0	46.13	4100.0	34.47

118

References

Air Products and Chemicals, Inc., 1987: Air Products Specialty Gases Catalog, Box 538, Allentown, PA 18105.

Briggs, G. A., 1973: Diffusion Estimation for Small Emissions. ATDL Rep. No. 79, ATDL, P.O. Box E, Oak Ridge, TN 37830.

Briggs, G. A., 1975: Plume Rise Predictions, in *Lectures on Air Pollution and Environmental Impact Analysis*, AMS, 59–111.

Ermak, D. L. and S. T. Chan, 1985: A Study of Heavy Gas Effects on the Atmospheric Dispersion of Dense Gases. *Proc. 15th Annu. Int. Tech. Meeting on Air Pollution Modeling and Its Application*, UCRL-92494, LLNL, Livermore, CA 94550.

Fauske, H. K. and M. Epstein, 1987: Source Term Considerations in Connection with Chemical Accidents and Vapor Cloud Modeling. *Proc. Int. Conf. on Vapor Cloud Modeling*, CCPS/AIChE, 345 E. 47th St., New York, NY 10017, 251–273.

Fleischer, M. T., 1980: SPILLS An Evaporation/Air Dispersion Model for Chemical Spills on Land, Shell Development Company, Houston, TX.

Golder, D., 1972: Relations among Stability Parameters in the Surface Layer. *Bound. Lay. Meteorol.* **3**, 47–58.

Goldwire, H. C., T. G. McRae, G. W. Johnson, D. L. Hipple, R. P. Koopman, J. W. McClure, L. K. Morris, and R. T. Cederwall, 1985: Desert Tortoise Series Data Report—1983 Pressurized Ammonia Spills. UCID-20562, Lawrence Livermore National Laboratory, Livermore, CA 94550.

Hanna, S. R., G. A. Briggs, and R. P. Hosker, 1982: *Handbook on Atmospheric Diffusion*. DOE/TIC-11223, DOE, 102 pp.

Hanna, S. R. and P. J. Drivas, 1987: *Guidelines for Use of Vapor Cloud Dispersion Models*, CCPS/AIChE, 345 E. 47th St., New York, NY 10017, 177 pp.

Havens, J., 1987: A Dispersion Model for Elevated Dense Gas Jet Chemical Releases. University of Arkansas, Fayetteville, AR 72701.

Havens, J. A. and T. O. Spicer, 1985: Development of an Atmospheric Dispersion Model for Heavier-Than-Air Gas Mixtures. Coast Guard Report CG-D-23-80, USCG HQ, Washington.

Hoot, T. G., R. N. Meroney, and J. A. Peterka, 1973: Wind Tunnel Tests of Negatively Buoyant Plumes, CER73-74TGH-RNM-JAP-13, Colorado State Univ., Fort Collins, CO.

Lyman, W. J., W. F. Reehl, and D. H. Rosenblatt, 1982: *Handbook of Chemical Property Estimation Methods: Environmental Behavior of Organic Chemicals*. McGraw-Hill, New York.

Ooms, G., A. P. Mahieu, and F. Zelis, 1974: The Plume Path of Vent Gases Heavier Than

Air. *First Int. Symp. on Loss Prevention and Safety Promotion in the Process Industries*. The Hague.

Perry, R. H. and D. Green, 1984: *Perry's Chemical Engineers' Handbook*, 6th ed. McGraw-Hill, New York.

Reid, R. C., J. M. Prausnitz, and T. K. Sherwood, 1977: *The Properties of Gases and Liquids*, 3rd ed. McGraw-Hill, New York.

Index